The Mystery Animals of
# PENNSYLVANIA

**ANDREW GABLE**

Edited by Corinna Downes and Richard Freeman
Typeset by Jonathan Downes,
Cover and Layout by SPiderKaT for CFZ Communications
Using Microsoft Word 2000, Microsoft Publisher 2000, Adobe Photoshop CS.

First published in Great Britain by CFZ Press

**CFZ Press
Myrtle Cottage
Woolsery
Bideford
North Devon
EX39 5QR**

© CFZ MMXII

All rights reserved. Without limiting the rights under copyright reserved above, no part of this publication may be reproduced, stored in or introduced into a retrieval system, or transmitted, in any form of by any means (electronic, mechanical, photocopying, recording or otherwise), without the prior written permission of both the copyright owners and the publishers of this book.

# ISBN: 978-1-905723-95-9

# CONTENTS

| | | |
|---|---|---|
| 1/ | Giant Skeletons | 5 |
| 2/ | Native American Legendry | 11 |
| 3/ | Monsters of New Sweden | 25 |
| 4/ | The Eternal Hunter and Others | 29 |
| 5/ | The Bird of Happy Omen | 41 |
| 6/ | Animal Superstitions | 45 |
| 7/ | Hex and Violence | 47 |
| 8/ | Bizarre Tales of the Animal World | 59 |
| 9/ | The Broad Top Snake | 69 |
| 10/ | Water Monsters | 79 |
| 11/ | The Tinicum Swamp Cat | 95 |
| 12/ | The Yardley Yeti | 99 |
| 13/ | The Ape-Boy of Chester | 101 |
| 14/ | The Jersey Devil | 109 |
| 15/ | Montie the Monster | 123 |
| 16/ | Thunderbirds of the Black Forest | 129 |
| 17/ | Werewolves | 141 |
| 18/ | Black Dogs and Phantom Hounds | 149 |
| 19/ | Bigfoot in Pennsylvania | 163 |
| 20/ | The Rest of the Weird | 177 |

Appendix: The Great Circus Train Wreck of 1893
        (*Huntingdon Daily News*, May 29, 1943)    193
Bibliography and Acknowledgements    201

*For my wife, Mindi*

# Chapter One
## Giant Skeletons

The valley of the Susquehanna River extends from its headwaters in New York state southward across Pennsylvania until it empties into the Chesapeake Bay near Havre de Grace, Maryland. The entire valley in pre-Colonial time was inhabited by a powerful tribe known as the Susquehannocks, from whom the name of the river is derived. Other Native American tribes of Pennsylvania, the Andaste, Conestoga and Minqua, were merely alternate names of this tribe. They spoke the language of the Iroquois tribes, though they were traditional enemies of that alliance of tribes which bedeviled settlers of western New York state. The territory of the Susquehannocks was not confined to the regions along this river, however, as remains of the tribe have been retrieved from the "46HM73" site in Hampshire County, West Virginia along the Potomac.

As long ago as 1608, Captain John Smith of England described the impressive stature of the Susquehannocks, who were also present throughout much of the Chesapeake Bay region. His account, though riddled with the creative and archaic spellings common to the era, casts the Susquehannocks as paragons of the "noble savage" conceit, barbarically-dressed and physically imposing, but yet peaceable (however, this image does not tally with his earlier mention of "warlike Iroquois").

> "Such great and well proportioned men are seldome seene, for they seemed like Giants to the English... yet seemed of an honest and simple disposition... for their language it may well beseeme their proportions, sounding from them as a voyce in a vault. Their attire is the skinnes of beares, and wolves, and some have cassocks made of beares heads and skinnes... One had the head of a wolfe hanging in a chaine for a Jewell... [they were armed] with Bowes, Arrows and Clubs, suitable to their greatnesse... The picture of the greatest of them is signified in the Mappe, the calfe of whose leg was three-quarters of a yard about, and all the rest of his limbes so answerable to that proportion, that he seemed the goddliest man we ever beheld."

Several villages of the Susquehannocks were described by Captain Smith. Sasquesahanough was at the mouth of the Octoraro Creek south of Conowingo, Maryland; Quadroque was near the town of Washington Boro (Lancaster County); and Tesinigh was at present day Falmouth (also in Lancaster County), near the Three Mile Island nuclear plant.

In the same account of his travels, Smith describes a tribe of cannibals, "eating the flesh of their victims boiled," called the Andastes. As already described, Andastes was merely the name given to the Susquehannocks by the French (George Alsop, writing in 1666, also alludes to the Susquehannock habit of cannibalizing their war captives). In 1763, during Pontiac's Rebellion in the Ohio Territory, the Paxton Boys gang in Pennsylvania murdered several Susquehannocks living near present-day Millersville (Lancaster County), claiming that they had been treating with hostile tribes. In response, Governor John Penn relocated the remainder of the tribe to the Lancaster city prison for their safety until the gang could be apprehended. On December 27, the vigilante gang infiltrated the prison and murdered twelve of the fourteen survivors. The scene at the prison must have been truly grotesque and not out of place for a slasher film, as shown in this gruesome description by Lancaster native William Henry:

> "I saw a number of people running down street towards the gaol, which enticed me and other lads to follow them. At about sixty or eighty yards from the gaol, we met from twenty-five to thirty men, well mounted on horses, and with rifles, tomahawks, and scalping knives, equipped for murder. I ran into the prison yard, and there, O what a horrid sight presented itself to my view! Near the back door of the prison, lay an old Indian and his squaw (wife), particularly well known and esteemed by the people of the town, on account of his placid and friendly conduct. His name was Will Sock; across him and his squaw lay two children, of about the age of three years, whose heads were split with the tomahawk, and their scalps all taken off. Towards the middle of the gaol yard, along the west side of the wall, lay a stout Indian, whom I particularly noticed to have been shot in the breast, his legs were chopped with the tomahawk, his hands cut off, and finally a rifle ball discharged in his mouth; so that his head was blown to atoms, and the brains were splashed against, and yet hanging to the wall, for three or four feet around. This man's hands and feet had also been chopped off with a tomahawk. In this manner lay the whole of them, men, women and children, spread about the prison yard: shot, scalped, hacked and cut to pieces."

Only two members of the Conestoga Susquehannocks survived the Paxton Boys massacre; a reward was offered for the capture of the gang's leaders by Governor Penn, but although the murders were widely condemned and a criticism of the gang written by Benjamin Franklin himself (who said "they were a drunken, debauch'd, insolent, quarrelsome Crew"), none were ever apprehended. The prison, by the way, where this took place is not the current Lancaster prison with its castle-like façade, but is no longer in existence. It stood on the site of the current-day Fulton Opera House (which is also supposedly haunted, although it seems that the ghosts have nothing to do with the massacre. Rather, they are the phantoms reported from nearly every theater worldwide) and the gate to the old prison is preserved in the foundation wall of the opera house.

The stature of the Susquehannocks has already been alluded to in Smith's account. But this is not the typical exaggeration common to so many traveler's tales: skeletons recovered from the 46HM73 site in West Virginia, also described earlier, confirm Smith's descriptions of the Susquehannocks as "Giants". When a railroad bridge was being constructed near Conowingo, Maryland (another site known to be inhabited by Susquehannocks) large skeletons were also discovered: Johnston (1881) says merely that they were "the remains of persons of extraordinary size, seem[ing] in some measure to confirm [Smith's] account", and no more precise measurement is available. Again, George Alsop says of the Susquehannocks that "the men for the most part [were] seven feet in latitude."

An article published in *American Antiquarian* in 1885 described the well-preserved remains of a man fully 7'2" in height, crowned with a copper circlet, found in a burial mound near Gasterville. The existence – or lack thereof – of any town called Gasterville proved to be problematic, until I discovered that the events described in the article appeared in an item appearing in the *Titusville Herald* on May 6, 1884. As revealed therein, the discovery of the giant remains took place not in Gasterville, Pennsylvania but in Cartersville, Georgia. A second mound mentioned in the *American Antiquarian* article as having been excavated concurrently – in Barton County – was revealed to have been a reference to Bartow County, Georgia.

In 1822, a cellar was being dug on the property of General McKean in Burlington Township, Bradford County, when a sarcophagus was discovered. The vault was opened and a skeleton 8'2" in height was discovered. It was noted that there were 10 other sarcophagi in the vicinity, but the contents of none of those were given.

In his *History of Waverly, New York* (1943) Charles L. Albertson states that in 1897, when the Spalding Memorial Library was being built in nearby Athens, in Bradford County (Pennsylvania) "they exhumed a skeleton about eight feet in height". A newspaper article from the time also states that a thighbone was found on the farm of Charles Hooker at Mount Pisgah, again in Bradford County, in about 1850. A local physician, Dr. Theodore Wilder, examined the bone and calculated that the man from whom it came must have been upwards of 7 feet tall.

The Eries (the tribe was also known to the French as *Nation du Chat*) lived in the northwestern portion of the state, as should come as little surprise given the name. They spoke an Iroquoian language like the Susquehannocks, though also like that tribe they were at war with the Iroquois proper; and, in fact, it was the Iroquois which exterminated the Erie tribe. Also like the Susquehannocks, great size was apparently far from unheard of among the Eries.

In 1820, Dr. Albert Thayer examined some bones pulled from a mound near Erie and declared that they "indicated a race of beings of immense size." Another discovery came sometime around 1875, when a skeleton between 8 and 10 feet tall was discovered in Conneaut Township.

W.H. Scoville excavated an Indian mound at Ellisburg, in Potter County, in 1886 and discovered the skeleton of a man nearly 8 feet tall; only a few years later, the skeleton of a man 8 feet tall was found beneath an oak tree at a mound on the property of John Pomeroy at Albion (Erie County). Even further afield, a skeleton between 8 and 9 feet tall was discovered in another mound, this time near Greensburg (Westmoreland County), in 1921.

The Erie giants are possibly referenced in the dubious *Walam Olum* or Red Score, originally published in the 1820s. Formerly thought to have been derived from sacred Lenape writings, the *Walam Olum* is now generally thought of as a complete fabrication, penned by an eccentric professor from Lexington, Kentucky named Constantine S. Rafinesque. It is thought that he executed such academic *faux pas* as finding Lenape words which fit the English meaning he wanted to give them, rather than in reverse as translation should go. Rafinesque's writings may also have been in part based on the accounts popularized by a contemporary scholar, John Heckewelder. Heckewelder wrote a manuscript called "*A Short Account of the Emigration of the Nation of Indians Calling Themselves Lenni Lenape (Since by the Whites Improperly Called Delaware Indians) As Related by Themselves*". Nothing short about that title, though. This work told of the fabled migration of the Lenape from the western United States, crossing the river they called the Namaessipu, thought to have been the Mississippi. Despite any fabrication of the whole of the works by either Heckewelder or Rafinesque, however, this tradition of migration is indeed present in Lenape mythology and is thought to possibly be a racial memory of the cross-country migrations of the original Native American populations originating in Asia.

The "horned skeletons" were found at the base of the Spanish Hill archaeological site in Sayre.

One of the hazards encountered along the way was a great battle between the Lenape and a group which lived along the Namaessipu, calling themselves the Talligewi. Heckewelder's descriptions of the Talligewi are of interest:

> "...hundreds of the Talligewi were slain... & who all went down the Ohio & Mississippi Rivers... Note, they say that the Talligewi, were a remarkable tall & stout People, & that There had been Giants among them – People of a much larger size than the tallest Of them (the Lenni Lenape) that they had built themselves angular fortifications – Or entrenched themselves, from which they could sally out, but were generally beaten – Both, of the fortifications & entrenchments, said to be built by them I have seen many, two Of which in particular were remarkable, the one being near the Mouth of the River Huron, Which emptieth in Lake St. Clair, on the North Side of that Lake, distance about 20 Miles N. East of Detroit... Walls, or banks of Earth had been regularly thrown up with a deep ditch on the outside & were on the Huron River, East of the Sandusky, & about 6 to 8 miles from Lake Erie... a number of large flat Mounds, in which the Indian Pilot said, hundreds of the slain Talligewi were buried..."

Heckewelder's writing abilities, this quote should make plain, were sorely lacking (all the more distressing as he was a learned man of his day). Regardless, it is surmised by some that the name Talligewi may be the basis for the word Allegheny, a river, mountain range, and county in western Pennsylvania. The notes Heckewelder made about the mounds in which the giants were buried is tantalizingly like certain stories told about the Lenape *mhúwe*, or cannibal giants, described in the next chapter.

So were the giants recorded among the Eries possibly a remnant of that tribe's having interbred with the conquered Talligewi? Were the giant Susquehannocks a similar interbreeding? Whether such great size was due to some genetic defect to which the Susquehannocks and Eries were prone or simply due to genetic variation is unknown. A case exists from along the Holly Creek in Kentucky in which the 'giant' individual was clearly deformed.

But how are these Natives of Unusual Size (apologies to *The Princess Bride* for that groaner) relevant to the modern anomalist? The story of the horned giants, whose skeletons were discovered by G.P. Donehoo, A.B. Skinner and W.K. Morehead near Sayre, Bradford County in 1916, is a well-known one in the annals of Pennsylvania's anomalies and has been told and retold for years. However, the truth of this matter has been known since 1921. In that year Louise Welles Murray, resident of the property on which the excavations took place, wrote an account of the dig (although, as we've seen, the 7 foot stature attributed to the skeletons was likely not an exaggeration):

> "While the writer was present one of the men working a grave exclaimed, "There are horns over his head!" Mr. Skinner said that indicated chieftainship... A passing visitor, however, heard the exclamation and attempted to verify it by interrogating a fun-loving Maine workman, and the story grew and was printed from coast to coast that one or more skulls had been found with horns growing from the forehead!"

# Chapter Two
## Native American Legendry

Pennsylvania has had a number of Native tribes live within its borders over the years. Most notable among those have been the Lenape, also known as the Delaware (these are the Indians pictured in the famous painting of William Penn meeting with their chief); the Susquehannock, already described in the last chapter; and the Seneca, who were a part of the Iroquois Confederacy, an alliance of a number of smaller tribes which dominated New York state and were engaged in a number of wars with colonists there. The Seneca later moved into western Pennsylvania and of all the tribes had the most contact with the Lenape (a fact attested to in their mythology).

The Piscataway tribe, dwelling in the regions along the Potomac River in Maryland (much of northern Maryland is a hotbed for Bigfoot sightings), venerated a spirit variously written as Okee or Ochre, as recorded by Mark Opsasnick and many writers before him. Okee seems to have been some sort of forest-dwelling wild man; Captain John Smith says that the Piscataways often assumed his image. Given the common origins of that tribe peoples as relatives of the Lenape, it is possible that Okee was similar to the spirit Mesingw.

The Lenape, referenced in Heckewelder's poorly-written treatise and Rafinesque's probable hoax, were an Algonkian-speaking tribe living all along the Delaware River and throughout the state of New Jersey. Over the past few centuries, the tribe has gradually been moved west to their current homeland in Oklahoma. Mesingw, also known to the Unami (one of the two major divisions of the Lenape) as Misinghalikun, was the "living solid face" or "mask being". He was a hunting spirit, a shepherd of the animals (deer in particular) who was often seen mounted on an elk. He was a wild, shaggy individual, with a mask, painted half-red and half-black, covering his face. Mesingw had a hooting cry. During the Big House Ceremony (a Lenape festival held in the autumn) it was customary for a member of the tribe to don a bear-skin robe and mask and impersonate the spirit.

> "After the dance is underway the Messinq [sic] comes from the darkness, jumps over the dancers, and dances between the other dancers and the fire. He makes some funny and queer gestures, kicks the fire, and then

> departs... The Devil Dance is what the white men call it, but... The Messinq does not represent an evil spirit, but is always considered a peacemaker."

Like Okee, Mesingw also was called "hideous" and a "devil" by Europeans, mainly because of his savage and uncouth appearance, rather than because of any malevolence on his part, although certain bits of folklore do seem to suggest that Mesingw was a vengeful, though not unjust, spirit.

Mesingw is also said to be the Lenape's version of the Sasquatch, based primarily upon his appearance. Many mention the vocalizations ("ho-ho-ho") recorded for the spirit entity as also reminiscent of that wildman. Any supposed similarity of Mesingw with Bigfoot, however, seems to originate in misunderstood references to the dancers of the Big House ceremony, and though there are references in various folktales to a number of other "mask beings," there is absolutely no evidence to suggest that the Lenape considered Mesingw or these mask beings to have any sort of reality beyond that of a semi-divine being.

Mesingw is also mentioned in connection with the Jersey Devil of the New Jersey Pine Barrens. This, though, is a somewhat interesting notion considering Mesingw's unfair label of "devil" and a perception among some Native Americans that the Jersey Devil is a beneficent force. And like Mesingw, the Jersey Devil is associated in folklore with storms.

In 1666, George Alsop wrote of the Susquehannocks that

> "[The Devil] is all the God they own or worship; and that more out of a slavish fear than any real reverence to his infernal or diabolical greatness, he forcing them to their obedience by his rough and rigid dealing with them often appearing visibly among them to their terror, bastinadoeing them, with cruel menaces even unto death and burning their fields of corn and houses that the relation thereof makes them tremble themselves when they talk of it...
>
> Once in four years they sacrifice a child to him in an acknowledgement of their firm obedience to all his devilish powers and his hellish commands... [the priests] oftimes raise great tempests when they have any weighty matter or design and by blustering storms divine his purpose."

Given that Thomas F. Gordon in his 1829 *History of Pennsylvania* states that the Susquehannocks were descended from the Wolf Clan of the Lenape and that "Dr. Shea also says... Alsop's view of the religion of the Susquehannocks is wrong – that they generally believed in a good deity," could it be that the Susquehannock "Devil" is an entity related to Mesingw, or possibly even Mesingw himself? In the Devil's "often appearing visibly among them," perhaps we have a reference to something similar to the Big House Ceremony and its Mesingw-suited dancers, and in the divining tempests a reference to Mesingw's association with storms.

It must be kept in mind that George Alsop was writing with a distinct pro-Maryland bias, and the colony of Maryland had a rocky relationship with the Susquehannocks.

In one Lenape tale since inherited by the Seneca (one of the tribes of the Iroquois Confederacy), a white stag known as *Ganyogowa* (great game) is a creature which "commands all the birds and animals," a sort of variant Mesingw. A ghostly white stag is reported from Shamong Township, New Jersey, in the heart of Devil country, where it is usually seen as a sort of portentous apparition (Shamong being a Lenape word meaning "place of the horn"; also it should be noted that the similarly derived Chemung River flows past Sayre, Pennsylvania, location of the discovery of the "horned" Susquehannock skulls described in the previous chapter). If the white stag can stand in for Mesingw, it is then interesting that the Jersey Devil, too, has a role in some folklore as a crisis apparition of sorts.

Many other beings featured in the mythology of the Lenape, some of which play prominent roles in the lore of other Indian tribes, as well. The *mëxaxkuk* or *w'axkōk* was the horned serpent (a monster which appears in the folklore of many other tribes as well); a water monster that dwelt among rocky piles in the midst of the rivers. This was a spirit who controlled plague in some tales, and it seemed to be a sort of fertility spirit as well, either granting young men the ability to be irresistible to women or, occasionally, taking the form of a man itself and impregnating a Lenape girl, causing her to give birth to tiny snakes. The *wewtamíwes* or merman was a similar water monster, but despite the name it could be male or female. It has a lecherous nature like the horned serpent, and is also well-known as devouring Lenape that fall into its grasp.

Another familiar sort of creature was the thunderbird, known as *pléthoak*, *palésawak*, or *pethakhuwéyak*. Although usually appearing as a human, and in this form referred to as Thunders, or sometime Thunder Boys, they are nonetheless referred to as birds – usually eagles, sometimes game birds such as turkey or grouse. It was also believed that the thunderbirds could bestow

**The controversial Lenape Stone, found near Wycombe in 1872.**

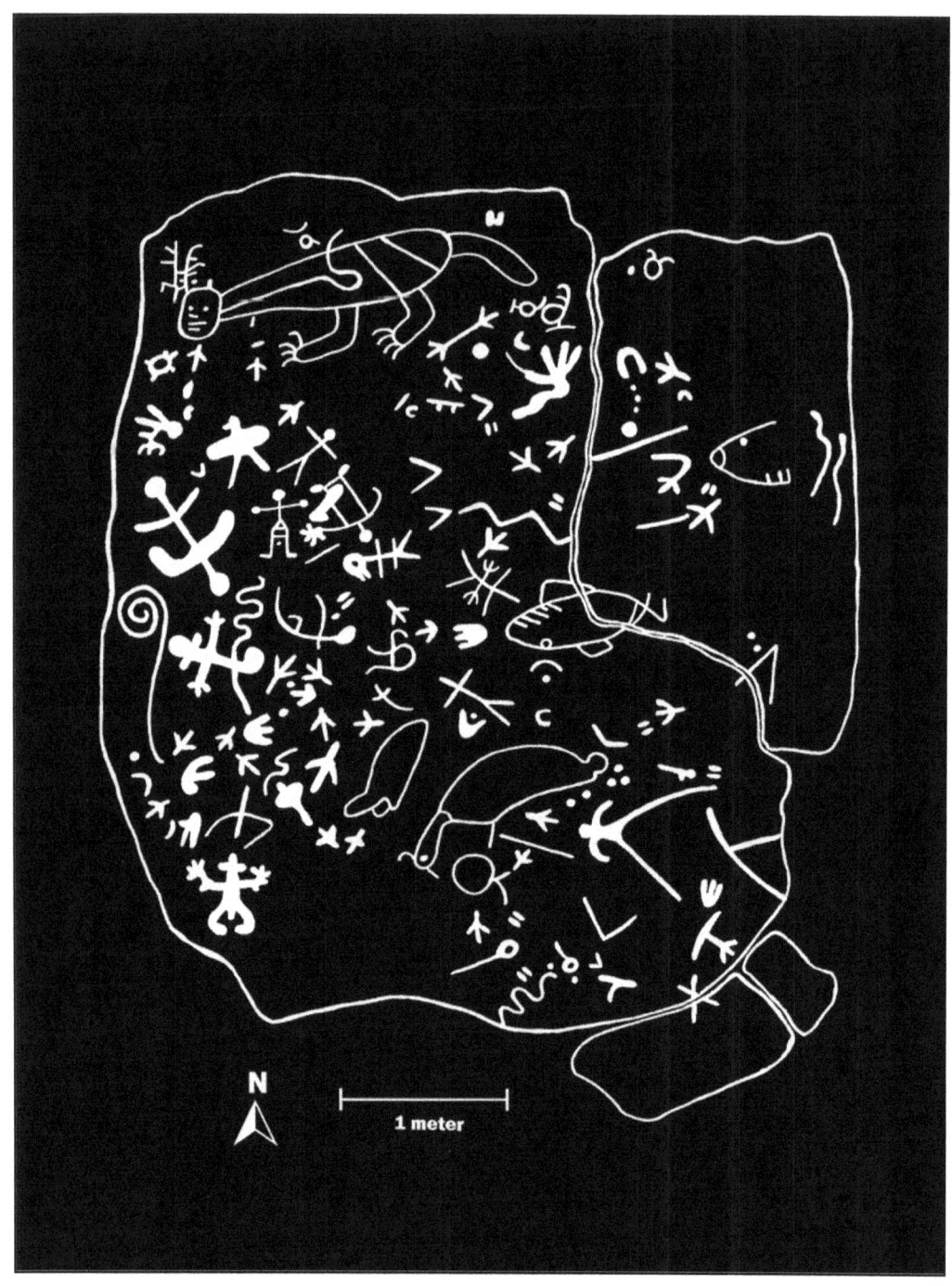

Petroglyphs found at the Parker's Landing site, with the "water panther" at the top.

their power upon certain individuals. Many other Native American tribes have traditions stating that the thunderbird is a giant transformed by wearing a feathered cloak. The spirits are known for an animosity with the *mëxaxkuk* or horned serpent.

Yet another being was the *mhúwe*, also called *mamuui* or *málew*. These were giants, cannibals; and possibly there is some connection to the Susquehannocks and their massive size and the cannibalistic habits bestowed upon the Andastes population. Some tales give their origin as ones who were caught in snowstorms and turned to cannibalism, and also mention that they live in the north; this variant of the tale seems to have much in common with the otherwise Algonkian tales of the anthrophagous giants *wendigo* and *chenoo*. Individuals could be transformed back into a human by being forced to eat food other than human flesh. Other tales indicate that the cannibals roll in tar and gravel, becoming armored and immune to arrows (much like the Stone Coats of Iroquois mythology). Despite their fearsome nature, however, the giants are also stupid and easily tricked (in one folktale, some hunters escaped from a giant who assumed he couldn't cross a river, having measured its length rather than its depth). Interestingly, one tale also notes of the *mhúwe* that "[i]t is they who are buried in Indian mounds."

The *mhúwe* is probably much the same as the whistling bear-faced giant called *gougou* about which Samuel de Champlain was told by the Micmac tribe (more Lenape relatives) in 1603. It is also called *gugwes*, which in addition to probably being the derivation of Lovecraft's name for certain subterranean, pink-eyed giants of the Dreamlands is more similar to the Lenape name (also doubtless the same as the Penobscot *kiwakwe*).

At one time, say the Lenape, the squirrel was a gigantic monster, ravenously devouring everything that came his way, including a human being. But the squirrel was cursed by the Creator for eating the human (apparently, the callous Creator didn't care about all the animals he ate) to become small, and carry the man's hand around with him, under his arm, for all eternity. One almost wonders whether the 'giant squirrel' could have been memories of the gigantic ground sloths of prehistoric days.

A creature which I can find no direct mention of in Lenape lore, but which definitely did exist among the lore of the Algonkian tribes (more specifically the Ojibwa, better known, if incorrectly, as Chippewa) was called *bakaak*, or "skeleton". This was a skeletal figure with glowing red eyes, which roamed the forests bedecked in the garb of a hunter. When it found an Indian warrior, it would assail him with a club and invisible arrows; and when slain, it would eat his liver (liver-eating is a typical feature of Native American ogres). Bakaak was reputed to have been a murderer cursed by the gods. He appears in *"The Song of Hiawatha"* as *pau'guk*, and is made a spirit of death in Iroquois lore.

Though I know of no direct link between the *bakaak* and more familiar man-eaters of Indian lore, we must note that the traditional appearance of the *wendigo* among the Ojibwa is as a being

> ...gaunt to the point of emaciation, its desiccated skin pulled tautly over its bones. With its bones pushing out against its skin, its complexion the ash gray of death, and its eyes pushed back deep into their sockets, the Wendigo looked like a gaunt skeleton recently disinterred from the grave.

Given the skeletal appearance of the more familiar creature, it seems at least possible some cross-contamination of the two myths existed.

But lest the reader think I'm mentioning a creature for little reason except a certain undeniable charm it possesses, a local legend exists around Glen Lyon (Luzerne County, just outside of the Wilkes-Barre/Scranton metropolitan area) of a ghost called "Bagunk" which haunts the cemeteries of the town. It's been noted that Bagunk is a name whose derivation is a mystery in itself: given that one of the alternate names of this skeletal hunter was *baguck*, one wonders whether that's the derivation. Of course, I'm at a loss as to why the name of a skeletal liver-eater would have become associated with a typical graveyard phantom: unless there is more to the Bagunk legends then we know.

Another spectral legend of a skeletal kind is the Ghost of Longswamp or *Schpuk von Langeschamm*, whose story was told by poet John Birmelin. The ghost was supposedly the specter of a cruel ironmaster working at the Mary Ann Furnace. So cruel, in fact, that he killed a furnace worker by tossing them into the blazing interior of the furnace. After death, he was cursed to walk the corner of eastern Berks County known as Longswamp as a bony skeleton.

In the story of the 'schpuk', we can see clearly the influence of the *bakaak* legend: a murderer, sentenced to exist eternally as the skeletal remains of what he once was. And as will become more clear in Chapter Five, a good deal of supernatural lore centers around the iron-smelting industry in Pennsylvania. But the story doesn't end there!

The rather bossy spirit demanded that the living residents of Longswamp pay him tribute to get him to stop raiding their farms and generally causing all sorts of ghostly havoc. And in the grand old tradition of amazingly random things in mythology, to quell the chaos he caused he demanded to be presented with a broom, a pair of shoes, and a straw hat. Exactly why the ghost needed those things, I don't know, and I likewise don't know why it was apparently so difficult for the residents to acquire the items.

Rather than give him these three relatively commonplace items, the residents chose to summon a priest who dwelt on Diehl's Head Mountain or *Diehlekopp* to exorcise the spirit. He was captured near Henningsville and carried to the top of the mountain where the priest drew a banishing circle around his form, muttered some incantations, and banished the skeletal *schpuk* to Hell. You have to wonder whether this guy was really a priest, because talk of magic circles and incantations sounds like rather un-priestlike behavior.

The fact that he was captured, and carried, makes it seem that this specter was, well, not a ghost. The *schpuk* seems to have been a little less bloodthirsty than the hunter-slaying, liver-

## The Mystery Animals of Pennsylvania

eating *bakaak*, but I'm sure they had to clean the tale up for all the little kiddies of Longswamp.

A number of Thunderbirds are depicted among other forms in the petroglyphs or rock-carvings (see below) found on Big and Little Indian Rocks, near Safe Harbor (Lancaster County), likely of Susquehannock origin. Some of these petroglyphs were removed and sent to the State Museum in Harrisburg to avoid being inundated when the Safe Harbor Dam was constructed.

Making our way into the western portion of the state, a number of petroglyphs exist that may have some relevance to unknown creatures. A number of rocks near Parker (Clarion County), on the banks of the Allegheny River, are bedecked with carvings, most likely the work of the Seneca tribe. The carvings are a mixed bag, depicting all manner of animal life (among them a turkey, a turtle, a fish and what seems to be a mountain lion) and, on a rock designated Group 6 by researchers from the Clarion County Historical Society, a cluster of human figures and an "antlered panther," which is doubtlessly a depiction of the spirit known as a "water

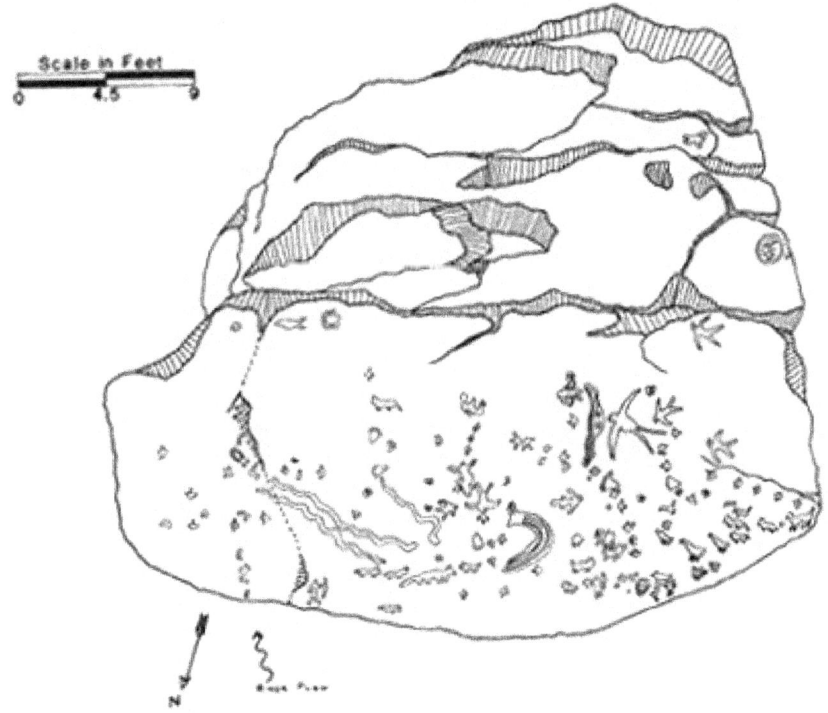

FIGURE 3. Little Indian Rock petroglyphs (Source: Cadzow 1934)

panther", which figures in the beliefs of many Native American tribes, particularly in the Northeast. A carving found at the Indian God Rock petroglyph site 15 miles away may also be an attempt to depict a water panther.

The water panther was usually depicted as – you guessed it – a panther, with a humanlike face, antlers, and a long tail, which had a fin of some type, which was so long that it curled up and over the "panther." It was believed that the panther stayed in the river, and that the thrashing of his tail caused rapids in the water. It was also believed by certain tribes that he could change himself into a shooting star and in this form fly between bodies of water. Many people believe that the infamous *piasa* rock painting in Alton, Illinois, was originally a depiction of a water panther (and, indeed, the wingless *piasa* depicted on a 1678 map of the region is uncannily water panther-like).

Besides the earlier-described *Ganyogowa*, there were a few other mythical beings in the lore of the Seneca tribe that are of interest. A few Seneca tales speak of a cannibal named *Ongweias*, who was the brother of the creature *Dagwanoeyent* (big head), which has also been mentioned in cryptozoological context as a giant owl. It seems, however, that this creature, which appeared as a floating head, was the spirit-being representing the whirlwind or storm. Conflictingly though, some of the tales speak of people who have become *ongweias*, using the term in the plural. The *gougou/gugwes* is often spoken of in connection with Bigfoot stories; the Seneca also had tales of a forest-dwelling humanoid being called *gadjiqsa* (husk false face) which often imparted upon the discoverer knowledge of how to defeat some monster.

The strangest being of all those described in Seneca lore, though, has to be the being called the Skin Man. An inflated human skin emptied of its contents, Skin Man, or *Hadjoqdja*, hung from the trees in the forest near a strawberry patch, which he protected.

> The Penobscot belief was that when a common black bear had once eaten human flesh, the bear's hair or fur would fall out entirely, and the bear would become naked and hairless. Once having tasted human flesh, the bear henceforth became dangerous and would attack other human beings. At this time the bear became the possessor of magic or supernatural power, so that he could charm and subdue those who chanced to meet him in the woods.

Although this passage describes a being in the mythology of the Penobscot tribe of Maine which was called the *Wa'kwekkeh* or Great Hairless Bear, as a tribe of the Algonkian language group the Lenape were the distant cousins of the Penobscots. They have legends of a nearly identical creature called *ya'kwahe* or *Amangachktlát* (a rather insulting name meaning Big Rump).

> There is likewise a kind of bear, much larger than the common bear, with much hair on the legs, but little on the bodies, which appear quite smooth. The Indians call it the king of bears, for they have found by experience that many bears will willingly follow it. While all bears are carnivorous... this kind of bear is particularly voracious...In more northerly

regions, as, e.g., in the country of the Mingoes [the Lenape name for the Iroquois], these are more frequently found and they have killed many Indians.

The Iroquoian tribes were even more insulting, upgrading the adjective and naming it Huge Rump Bear, or *Ahamagachktiâtmechquá*. The bear was long-bodied and massive, and totally lacked fur except for a whitish patch on the chest.

Another monstrous bear among the Iroquois was *Nyagwaihegowa*, the Ancient of Bears, also hairless save for a patch on the end of its tail. The hairless bear killed anyone who saw its footprints until a young boy shot it in the foot with an arrow (its heart being protected from harm by its ribs, which were one solid bony plate) and it died; the gods were fond of sending him to punish the wicked. Like with other mythological creatures, some Seneca myths refer to a number of Ancients. The Senecas also have a legend about a similar creature called *Dzainosgowa* (Ancient of Lizards), a gigantic blue lizard that could summon storms, and make night of day. It hopped from tree to tree, and was finally killed by a meteor. It was much like the hairless bear in personality.

Another variant of the naked bear story is found among the far-flung Gwich'in tribe, in Alaska. Although unrelated to either the Lenape or the Iroquois, they nevertheless have a tale that an especially voracious bear will roll in the snow and coat himself in an impenetrable icy crust (these ice bears are called *ĕ'a't'oánh*, and their parallels with the Stone Coats of Iroquois myth and some tales of the Lenape *mhúwe* giants are undeniable). The hairless bear, then, seems to be an ursine form of the familiar *wendigo* myth common to many northern tribes: the man who becomes a cannibal, becomes addicted to human flesh, and thereafter becomes an anthropophagous monster roaming the wilderness and eating travelers. But it seems also, given the descriptions, that the stories were perhaps originally descriptions of bears afflicted with mange or some other disorder. Increased aggression has been noted in animals with mange.

One of the tales of the Naked Bear noted by W.D. Strong, though, begins to throw another aspect into the myth. He says that the Naskapi tribe living in Labrador have stories of a carnivorous monster called *Kátcheetohúskw*, who is described as "very large, had a big head, large ears and teeth, and a long nose with which he hit people." This is clearly a very elephantine description: Strong was skeptical of this, however, as a legend of mammoths surviving in Alaska was derived not from any tradition of the Inuit people, but from drawings of mammoths which had been shown to the natives by whites years before. Questioning of Naskapi elders, however, revealed that this monster had always been described this way. The earlier-discussed Penobscots also have traditions of a number of "great animals with long teeth" at first mistaken for moving hills by the hero of the tale, Snowy Owl. Rivers were drying up because of their drinking. Since they were "so huge that when they lay down they could not get up", they slept by leaning on trees. Snowy Owl used this to his advantage, chopping down these trees and sharpening the stumps so that the animals were impaled whereupon they were all shot. Early Massachusetts clergyman Cotton Mather recounted a similar story, which he said took place in the Ohio region. Bizarrely, a nearly identical story of stiff-legged animals

sleeping against trees is recounted by Julius Caesar, and attributed to "elk", in the Hercynian Forest of Germany; and of course a parallel with the stiff-legged tall tale animal known as the hodag is also in order.

Many tribes of the American South shared the legends of the mastodon.

> A long time ago a being with a long nose came out of the ocean and began to kill people. It would root up trees with its nose to get at persons who sought refuge in the branches... It made its home in a piece of woods near Charenton [Louisiana, near New Orleans] and when guns were introduced the people went into the wood to kill the monster, but could not find it. When the elephant was seen it was thought to be the same creature, and was consequently called *Neka-cí ckamí*, 'Long-nosed spirit.'

(Another bit of linguistic synchronicity, the similarity of the last element to the Japanese *kami*, likewise meaning spirit, crops up here.)

In 1934, W.D. Strong noted that the Coushatta and Alabama tribes of the south "insist on translating man eater (*atipa tcobá*) as 'elephant'." The Atakapa tribe, also from Louisiana, have traditions that a huge monster was killed in a nearby waterway: this predated the discovery of a mammoth's skeleton in the Carancro Bayou in 1911. Thomas Jefferson himself took a great interest in the American mastodon, and in his *Notes on the State of Virginia* he recounted a Native American myth which he felt was about that ancient species of elephant.

> That in ancient times a herd of these tremendous animals came to the Big-bone licks, and began a universal destruction of the bear, deer, elks, buffaloes, and other animals... that the Great Man above, looking down and seeing this, was so enraged that he seized his lightning... and hurled bolts among them until the whole were slaughtered, except the big bull... [a lightning bolt] wounded him in the side; whereon, springing round, he bounded over the Ohio, over the Wabash, the Illinois, and finally over the Great Lakes, where he is living to this day.

Although it was later identified as Shawnee in origin, the tale originated with the Lenape. There are some references to a battle with the mastodons, presumably the same one described above, taking place at Buckwampum Hill (otherwise known as Ghost Hill) in northern Bucks County.

And so the tales circle back to Pennsylvania once more, but this time with the monstrous Naked Bear recast as a prehistoric proboscidean, and lest the readers wonder at the labeling of this herbivore as carnivorous in Jefferson's tale and elsewhere: is it any wonder that an animal with "teeth" several feet long was thought to devour men and animals alike? It has been noted that the valley of the Delaware was a hotbed of mastodon population. Skeletons recovered in New Jersey have had their stomach contents mostly intact, and one at Shawangunk, New York, even had bits of hair preserved.

The remains of mastodons can also be related in a roundabout way to at least some of the finds of giants noted above. The similarity of mastodon teeth to human's has been noted, and the bones of elephants have already been mistaken for those of giant humans in the past (the Greeks are thought to have based legends of the one-eyed Cyclops on discovered elephant skulls). Granted, a mastodon would inspire tales of a much larger giant than the 7 to 8 foot height usually noted, but a pygmy species could be to blame, at least partially.

In 1864 – so it's claimed anyway – Hilborne T. Cresson and W.L. de Suralt found an engraved whelk shell at Holly Oak, Delaware (on the shores of the Delaware River near Arden and just over the border from Pennsylvania) on which was depicted a long-bodied and short-legged elephantine animal with hair and straight tusks. The shell bore two holes in either end, and was likely worn as a pendant. The shell was first said to have been found in a peat bog on the grounds of a terminal of the Pennsylvania Railroad; later, in peat spread upon the field of a farmer. And the line was later said to be the Wilmington and Baltimore Railroad, not the Pennsylvania.

Cresson, a field assistant attached to the Peabody Museum in Massachusetts, and an amateur archaeologist with several digs in the Delaware Valley under his belt, didn't bring the Holly Oak shell to the public eye until 1889, 25 years after its alleged discovery. This, combined with the previously-discussed disagreements on where it was found, can lead anyone familiar with historical mysteries to begin to get some inklings of where the saga of the shell is going – recent dating of the artifact revealed that it was carved sometime in the 1880s, albeit on a shell genuinely acquired from an Indian site. Perhaps more hints to the artifact's spuriousness were given when very few, if any, archaeologists – even those well-acquainted with Cresson – referred to the shell.

In 1872, another pendant bearing a depiction of a mastodon was found in Bucks County. This one was stone, and was discovered by Bernard Z. Hansell on his father's farm near Wycombe. The rock, called the Lenape Stone, bore depictions of traditional motifs such as fish, turtles and birds. In the upper corner is a cluster of thunderbolts, a radiant sun, and most prominently, a mastodon. Unlike the Holly Oak depiction, this mastodon has more curved tusks. But like that carving, the Lenape Stone has likewise come under scrutiny. F.W. Putnam (who, perhaps coincidentally or perhaps not, was the superior of the H.T. Cresson later implicated in the Holly Oak find) was at first skeptical and then felt that the stone was genuine.

D.G. Brinton of Philadelphia's Academy of Natural Sciences summed up the opinions of the vast majority :

> "Certain artistic details, as the lightnings shooting in various directions from a central point (as from the hand of Jove), were also unknown to the art notions of the red race. The treatment of the sun as a face, with rays shooting from it, I also consider foreign to the pictography of the Delaware Indians, nor have I yet seen any specimens proved to be of their manufacture that present it. It is found, indeed, in Chippeway pictography, but there only in late examples.
>
> The execution of such imitations also usually betrays their origin. The lines on the Lenape Stone are obviously cut with a metal instrument, making clean incisions, deepest in the centre and tapering to points-quite different from the scratch of a flint point. Shrewder fabricators than the unknown author of this one make use of flint points. Some of the Western 'tablets' have been so inscribed. They may thus conceal their tools, but there are other resources for the archaeologist. The surface of all stones undergoes a certain chemical change on exposure to the air, which is called by the French term patine. In many varieties, as flints, jasper, and hard shales, this affords a decisive means of discriminating a modern from an ancient inscription or arrow-head. It requires the use of the microscope and some practice, but with these most of such impostures can be detected."

In his study of the Lenape Stone, though, Henry C. Mercer (an early proponent of the authenticity of the stone, despite the opinions of the majority of academia) notes that similar depictions of the sun do appear among a series of petroglyphs formerly located near Conowingo, Maryland, and also in another set found on several rocky islands near Safe Harbor (Lancaster County). This observation may suggest a Susquehannock, rather than Lenape, origin; this, of course, assumes that the stone is genuine – which is far from certain and, indeed, doubtful. Whatever its status, true artifact or modern hoax, the Lenape Stone currently resides at the Mercer Museum, an eclectic mixture of artifacts collected by Henry C. Mercer and housed in a monolithic structure in Doylestown (Bucks County).

# The Mystery Animals of Pennsylvania

**Stone carving of the face of Mesingw, found in Montgomery County.**

Tocks Island is located in the Delaware River directly below the Shawnee Mountain Ski Area in Monroe County (and not far from Columbia, New Jersey, former home of pioneering cryptozoologist Ivan T. Sanderson). In 1970, Herbert C. Kraft of Seton Hall University published a paper detailing the discovery of a pre-Lenape population of Palaeo-Indians, which was dated to 1720 BCE on the island.

The site was apparently also "recycled" and inhabited by the Lenape at a later date.

Another Warren County, New Jersey site along the Musconetcong River was also inhabited by the Palaeo-Indians. These two sites are located a good distance north of the Philadelphia area and so aren't directly relevant, but the Musconetcong site in particular (the site was nearly 10,000 years old) reveals that man and mastodon did co-exist in the region.

Interestingly, the BFRO has logged a Bigfoot report from along the Musconetcong River. The various tribes dwelling in Pennsylvania in prehistory were not the only group whose legendry may have relevance to reports of latter-day strangeness, as shall be seen.

# Chapter Three
## Monsters of New Sweden

In 1638, a Swedish ship called the *Kalmar Nyckel* (the Kingdom of Sweden, at the time, included present-day Finland, Estonia, and parts of Germany), captained by Peter Minuit, landed in what is today Wilmington, Delaware and founded the settlement of Fort Christina, named after the reigning Swedish monarch. Their claims extended northward through what is now Delaware County, and also across the river into Salem and Gloucester Counties in western New Jersey. Johannes J.H. Campanius (John Campanius Holm), a Lutheran priest, came to New Sweden in 1643. Campanius travelled up and down the Delaware, preaching to the Swedish settlers at their cabins. On his travels he encountered many strange and unusual creatures, which he cataloged in his writings.

One of these monsters was "a large and horrible serpent which is called a rattle snake, which has a head like that of a dog, and can bite off a man's leg as clear as if it had been hewn down with an axe!" It was noted that the cast-off skin of the snake had healing properties. The timber rattlesnake, also called the canebrake rattlesnake (*Crotalus horridus*) is found throughout central Pennsylvania and as far north as New England. The snake is usually about six feet long, and usually has alternating brown and black bands although some individuals are of a darker, almost black, coloration. Curiously enough, the black-tailed rattlesnake (*Crotalus molossus*), native to the southwestern United States and Mexico, is also known as the dog-faced (or dog-headed) rattlesnake.

Campanius also referred to animals he termed sea-spiders, which could grow to be "as large as tortoises". The sea-spiders had shells "of a kind of yellow horn. They have many feet, and their tails are half an ell long, and made like a three-edged saw, with which the hardest trees may be sawed down." It was recorded that swarms of the spiders were often found at *Spinnel Udden* (Spider Point), which is now Bombay Hook, near Leipsic, Delaware. Fantastic as these creatures sound from this description, it seems that what the Swedes called sea-spiders were nothing more than the common horseshoe crab (*Limulus polyphemus*). Huge numbers of horseshoe crabs emerge from the Delaware to breed on spring nights.

Another creature he described was a type of fish which he notes was called *manitto* by the Lenape. The *manitto* was possessed of monstrous teeth, and dived very deep into the water and blew water like a whale. It is also noted that there was only one spot in the river inhabited by the fish. The manitto may have been something similar to the physeter, a fanciful version of the sperm whale (*Physeter macrocephalus*), described in Rabelais' *Fourth Book* (1552):

> "Pantagruel spied afar off a huge monstrous physeter (a sort of whale, which some call a whirlpool), that came right upon us, neighing, snorting, raised above the waves higher than our main-tops, and spouting water all the way into the air before itself, like a large river falling from a mountain.
> 
> The physeter, coming between the ships and the galleons, threw water by whole tuns upon them, as if it had been the cataracts of the Nile in Ethiopia."

Two pygmy sperm whales (*Kogia breviceps*) were washed ashore at Woodland Beach, Delaware, just above Bombay Hook, in 2007 but had to be euthanized due to their injuries. Dolphins also are common residents of the lower reaches of the Delaware Bay.

But not all of the creatures encountered by Campanius and the other Swedes are so easily identified. A headless and monstrous animal called the *tarmfisk* (bowel-fish) was described in the account of one of Campanius' journeys as being:

> "...like a smooth rope, one-quarter of a yard in length, and four fingers thick, and somewhat bowed in the middle. At each of the four corners there runs out a small bowel three yards long, and thick as coarse twine. With two of these bowels they suck in their food, and with the two others eject it from them. They can put out these bowels at pleasure and draw them in again, so that they are entirely concealed; by which means they can move their bodies about as they like, which is truly wonderful to look upon. They are enclosed in a house or shell of brown horn."

At first blush, it would seem to be tempting to label the *tarmfisk* as being an eel; however, the Swedes would have likely been very familiar with this fish, plentiful as it is in Europe, and not described it as something odd. The only possibility I can think of would be a squid or some sort of similar animal, but even that seems not to fit the descriptions of the creature.

Whatever it may have been, it's not entirely unprecedented or unheard-of: the writer Olaus Magnus mentioned another vaguely similar animal called the *swamfisck* (hog-fish), which inhabited the seas around the motherland of these settlers and has a repulsive appearance not unlike that denizen of the Delaware described above.

> "Wherefore the first monster that comes, is of a round form, in Norway called Swamfisck, the greatest glutton of all other Sea-Monsters. For he is scarce satisfied, though he eat continually. He is said to have no distinct stomach: and so what he eats turns into the thickness of his body, that he appears nothing else than one Lump of Conjoyned Fat.

> He dilates and extends himself beyond measure, and when he can be extended no more, he easily casts out fishes by his mouth because he wants a neck as other fishes do. His mouth and belly are continued one to the other."

Magnus then goes on to describe a similar "rolling" behavior when threatened, much like the hedgehog or the armadillo, and also that it will consume itself at times when it is curled. It's interesting to wonder whether the *swamfisck*'s death by autocannibalism inspired the death scene of the spider-demon Ungoliant in J.R.R. Tolkien's *Silmarillion* (Tolkien, it is well-known, was influenced by Norse myth and legend).

Though behaviors are undoubtedly different in the possibly apocryphal accounts of the latter, both the *tarmfisk* and its possible analog the *swamfisck* bear some resemblance physically to one of the more repulsive, yet oddly endearing, members of the animal kingdom: a deep-sea dweller called the sea pig. Used as a generic name for a number of species (*Scotoplanes*) of sea cucumber, the sea pig is a bizarre little creature roughly the size of a human hand. Were it larger, it could surely exist in an H.P. Lovecraft story. It is globular in shape, possessed of four tentacular appendages, has a little cilia-fringed mouth at the "front" of its body, and a number of stubby little centipede-like legs.

Given this, perhaps the *tarmfisk* was representative of an animal seen by Scandinavian mariners, and was never intended as "reality". After all, as explorers of the world's first great wilderness (and for the most part, still the sea is still a wilderness!), the tales of sailors and fishermen the world over are rife with some of the strangest creatures ever conceived. And, presumably as a legacy of the Viking era and its focus on the sea, Scandinavian mythology is more rife than most with oceanic monsters.

# Chapter Four
## The Eternal Hunter and Others

No treatise on the strange creatures and other unexplained phenomena in Pennsylvania would be complete without an introduction to the most famous and iconic inhabitants of the state, the *Pennsilfaanisch Deitsch* (Pennsylvania Dutch, and despite the name they were Germans). Most of the Pennsylvania Germans originated from the area of the Palatinate (today's state of Rheinland-Pfalz). They arrived in this country in several waves, the first settling on the outskirts of Philadelphia in the 1680s and eventually founding Germantown. A second, much larger, wave of immigrants arrived in various American ports in the early 1700s in the wake of a number of wars which ravaged the Palatinate. Several of these, not liking the government of New York where they had initially settled, migrated southward and eventually settled in Berks County in 1723.

These Germans eventually began speaking a modified form of Low German (accounting for the sort-of-but-not-really nature of the snippet of the language above), and they also had carried a good deal of the legends of their homeland along with them. One of the more prominent of the German legends in Pennsylvania was that of the being termed *Ewich Yeager*, or *Ewige Jaeger*. The "eternal hunter" was usually a phantom, a product of the sort of hasty oath or ill-thought-out action which often spawns eternal wanderers in various European traditions, and which is quite similar to the tale of Peter Rugg, the New England farmer who swore he'd make it home by morning and ended up being cursed to wander forever. To be fair, the story of Peter Rugg is actually a complete work of fiction, but is often remembered as a folktale. In true melting-pot fashion, the German tales of the Wild Hunt or 'furious host' were recast as Pennsylvanian tales. Jakob Grimm's splendid 1835 book on Germanic myth and legend recounts the following:

> "In Lower Saxony and Westphalia this Wild Hunter is identified with a specific person, a certain semi-historic master of a hunt. The accounts of him vary. Westphalian traditions call him Hackelbärend, Hackelbernd, Hackelberg, Hackelblock. This Hackelbärend was a huntsman who went a hunting even on Sundays, for which desecration he was after death... banished into the air, and there with his hound he must hunt night and day, and never rest... The Low Saxon legend says, Hans

von Hackelnberg was chief master of the hounds to the Duke of Brunswick, and a mighty woodsman..."

This man von Hackelnberg had a dream in which he fought, and was conquered by, a great boar. He later came across the same creature in a forest one day, and fought with it. The boar injured von Hackelnberg's foot, which became infected, leading to his death. Grimm recounts the scene that transpired on his deathbed:

> "On his deathbed he would not hear a word about heaven, and to his minister's exhortations he replied: 'the Lord may keep his heaven, so he leave me my hunting;' whereupon the parson spoke: 'hunt then till the Day of Judgment!' which saying is fulfilled unto this day. A faint baying or yelping of hounds gives warning of his approach..."

These ones specifically were chosen as not only will they be clearly seen to be similar to the Eternal Hunter tales, but also that they originated in the same regions of Germany as most of the Pennsylvania Dutch.

Another legend which may have helped shape the image of the Eternal Hunter was that of *Mallt-y-Nos* or Matilda of the Night, a figure in Welsh legendry. A wicked Norman lady, she declared that she would "rather hunt than go to heaven."

The most famous version of the tale stateside originates in the iron-rich hills separating Lancaster and Lebanon Counties. In 1677, John Grubb emigrated from Cornwall in England to Wilmington, Delaware. An owner of a few tin mines in his home county, Grubb and his descendants were to become quite prominent in Pennsylvania's iron-smelting industry. In 1742, one of his sons, Peter, founded Cornwall Furnace in Lebanon County, named after his father's homeland; this is the site of the tale.

"*The Legend of the Hounds*", by George H. Boker, retells in verse form the story of a notoriously hedonistic and cruel ironmaster at one of the furnaces in the Cornwall area – the poem itself places the legend at Colebrook Furnace, while other versions of the tale place it at Cornwall Furnace proper. Likewise, the identity of the tale's antagonist, the cruel ironmaster, is disputed as well. It is often said to be Samuel Jacobs; other versions of the tale say that Robert Coleman, the wealthy tycoon who built Colebrook Furnace, was the "squire" of the poem. Yet others, those that move the story to Cornwall Furnace, have it that one of the Grubbs is to blame.

Regardless – while on a hunt in the forested hills around the furnace - the villainous ironmaster became enraged after his hounds refused to flush out a fox which he was tracking. The ironmaster viciously decreed that "if my dogs cannot hunt so well on earth, another hunt in hell!" – the entire pack of hounds was to be tossed into the furnace. This was done by the reluctant servants.

But the ironmaster soon descended into madness – he grew increasingly reclusive, took more and more to drink, and at times was seen acting as if he were petting and speaking to a dog

**Scranton's mysterious Woman in Black appeared near the Lackawanna Iron and Coal Company.**

none other could see. Servants in the household and employees at the furnace began to hear the baying of hounds. One day, the dying ironmaster had his bed moved to look out towards the smelter. He cried out once – the name of his favorite dog – and then claimed that flaming hellhounds were hunting his soul. With that, he died. A German huntsman employed on the grounds said that the ironmaster was now the Eternal Hunter, and would hunt with the hounds he betrayed for all eternity. To this day, the baying of hounds, the thunder of horses' hooves and the shouts of the ironmaster on the trail of his spectral quarry are heard in the Cornwall Hills.

A wonderfully tragic story – and one that could be easily dismissed as an especially melodramatic piece of fiction, were it not for the mention of the Eternal Hunter. The story bears a bit of similarity to the tale of the huntsman Dando – from Cornwall, by the way, a notable coincidence given from where the Grubbs emigrated – with the mentions of liquor as a sort of gateway-to-deviltry, and in fact it could be that it is some sort of cautionary tale (drink not too much, lest your soul be hounded for all eternity! Literally, in this case).

The Blue Mountains near Pine Grove (Schuylkill County) were also haunted by the figure of the Eternal Hunter. One summer in the early 1800s was a disastrous one for the valley; an extremely bad drought caused the creeks and waterways to dry up. This, in turn, caused the

local farms to fail which, in turn, caused the deer of the region to flee south over the ridge and seek greener pastures in Berks County.

During this time, a local man swore that he would find the deer and drive them back towards Pine Grove. As he set out for the southern ridges, he added an oath that if it took him forever to find the deer, so be it. He would even hunt through the sky, if necessary. It's unknown where exactly the man's fate befell him, or how it took shape, but nonetheless destiny caught up with him. Late at night in the autumn months of the year, it is said, the baying of hounds and the crack of gunshots can be heard in the darkened skies over the Blue Mountains.

Indian legends spoke of a spirit fox which frequented the forested ridges of the Yellow Mountains, near Galen Hall (Berks County). The spirit manifested itself as a red fox, but a freakishly large one, replete with glowing eyes and a tail which trailed sparks. The same legends also claimed that a great treasure in gold would come the way of anyone who hunted down and killed the spirit. In November of 1811, a hunt was mounted for the legendary fox by a local man, who made a solemn oath – on a Bible, no less, not merely to his fellows at the local tavern! – that he would catch the fox or roam the hills forever in pursuit of it.

The first blast from the hunter's rifle missed the fox completely; he catches the fox in his sights a second time and a bolt of lightning and peal of thunder – from a cloudless sky – causes the hunter's horse to rear up, throwing the hunter over a cliff to his fate. If you've at all paid attention to the other stories in this chapter, you can guess that fate – not merely his death, but transformation into the Yellow Mountains' variant of the Eternal Hunter.

Yet another Eternal Hunter tale originates in Fayette County, in the western part of the state (and, along with Westmoreland County, haunted by monstrous sightings of all types). In this one, an impoverished man living in the wilds found himself completely bankrupt after a gambling spree and unable to keep his dogs. The pauper tossed all the dogs into the Alliance Furnace, and now the phantasmal form of the man – along with his pack of hounds – roams the forests there.

A tale from Columbia (Lancaster County) seems similar. The Hunter himself doesn't figure in the story, but other elements of the familiar legend do. A man (whose name is unknown) was drinking at what is now the Fairview Tavern near Ironville. By some degree of coincidence, Ironville was a small settlement sprung up around a quarry owned by none other than the Grubbs, who have already come up in this section. Anyway, the drinker made his way out of the bar, swearing that "I'll be in Columbia by midnight, or I'll be in hell." Of course, as is the way with such tales, a riding accident on the way home claimed his life and the spirit of the rider (and his horse) is fated to roam the roads forever. At one point in the 1930s, a local man riding a horse on the Ironville Pike narrowly avoided a shotgun blast fired by some local who mistook him for the headless phantom. (I heard another version of this tale in my younger years, which had it that the *Union Hotel* in Columbia proper was the drinking establishment of the tale, and that it was a railroad spike lying on Lancaster Avenue that caused the rider's death.)

And with that we wrap up our discussion of the Eternal Hunter, but not of German folklore. Another piece of German folklore which seems to have made the trip to these shores are the *weisse frauen* or white women (*witten wijven* in Dutch). These were beautiful white-clad women, nearly always appearing in direct sunlight, which seem to have been associated with mountains. Oftentimes, the *weisse frau* bore a number of golden keys in her hand, with which to unlock a vault within which was stored some treasure. Jakob Grimm speculated that they were folk memories of the *ljósálfar*, the mound-dwelling elves which were ruled by the god Frey in Norse mythology. He also mentions in his discussion of the figures that in Denmark, they were thought to wear black, and that one encountered at Hennikendorf, Germany was half-white and half-black, suggesting the Norse goddess of the dead, Hela.

Such white women are mentioned often in ghostly encounters, not just in Pennsylvania but elsewhere. The white women are sometimes contrasted with black figures – these could be derived from the Danish versions, or could be an artifact of the duality of light and dark elves in Norse myth.

Adamstown (Lancaster County) is home to two spectral ladies called *die Weisse Fraa* and *die Schwarze Fraa*, the white lady and the black lady, named after the color of their dresses. The specters don't seem to acknowledge the living at all, but if you're brave enough to follow them, they'll lead you into the cemetery and then vanish.

I went to college in Lock Haven (Clinton County), and the theatrical building on the campus of the university is said to be haunted by a white woman and a black – well, it doesn't quite appear as a woman, but there is a marked antagonism between the white and the black entities. But more on that story in a later chapter.

A more maleficent woman in black was that reported from Scranton (Lackawanna County) in 1886-1887. She first appeared in Pine Brook, a mining area of the city, where she attacked several young girls returning to their homes from a party. Oh, and did I mention she could vanish into thin air? Well, she did, and she later made appearances in the nearby towns of Archbald and Carbondale. The black phantom supposedly disappeared into an old mine near the facilities of the Lackawanna Iron and Coal Company, and she was seen, eyes ablaze, on the Pittston Depot Bridge leading to West Pittston (Luzerne County).

By this time, all of Lackawanna and Luzerne Counties were in an uproar over the sightings of the black-clad lady. There was actually a bounty on the woman's head placed in Wilkes-Barre. Sightings of the mysterious "sable terror" were soon reported from Sebastopol (just outside Pittston), and later still from the vicinity of Nanticoke. Soon the Dark Lady of Scranton became more violent, assaulting several individuals – or, as the *New York Times* supposed on January 7, 1887, "the hallucination which was at first harmless has become serious by affording unscrupulous and criminally disposed persons an opportunity to do their wicked work under the mask of the Woman in Black." The hysteria had reached such a point that "it is not safe for a lady dressed in black to be out after dark." But after this story, the panic subsided and the specter – whoever or whatever she was – disappeared.

The "sable terror" known as the Woman in Black haunted the coal miners of Scranton.

A parallel to the nefarious deeds of the English bogeyman known as Springheeled Jack would be in order here, obviously, and this mysterious female specter should be classed alongside him in the annals of phantom attackers. Except that it was a 'she' – but in this case, one should keep in mind the similarly gender-bending nature of the Mad Gasser of Mattoon and others of the phantom attacker ilk; witnesses to the Mad Gasser, for instance, reported usually a man, at times a woman, at times a woman in man's clothing, and at times a man in woman's clothing.

Sometime in the early years of the twentieth century, Mary Ann Litzenberger was walking with her sisters along a country lane in Buchtown, a section of Manheim (Lancaster County) near what is today the Manheim Auto Auction. The following was her experience:

> "To my astonishment as I gazed across the field nearby, I saw the ghost of a stark-naked man rise up from a fence corner and slowly walk across the field, not looking right or left, but having a worried look on his face.

The Pittston Depot Bridge, another haunt of the Woman in Black.

> I yelled with all my might to my sisters, saying in Dutch, 'A nockisher mon! Don't you see him?' But they insisted they didn't see anything of the kind."

Later, Mary Ann was to ask a neighbor about the naked man.

> "She informed me that there really was such a spook, that she had seen it too, at different times. She explained that there had been a dispute about the correct line of property, and the ghost of the farmer-owner verified it by stepping it off where it should be."

Why the specter of the land-owner was nude, we might never know… and we might be better off for not knowing. Perhaps in the Afterlife, they force you to disrobe for some horrible reason known only to the dead. Unusual though it may seem, this is far from an isolated occurrence, however.

W.J. Hoffman reports two more stories of similar phantoms.

The first concerns a miser living in Tulpehocken Township (Berks County). He moved the boundary-markers between his and his neighbor's property, so that he could encroach onto his neighbor's farmland. After the old man died, the farm became well-known in the vicinity as a haunted spot, and many of the other residents of the township flocked to the fields to watch the poor old man's ghost flit aimlessly around. If you ask me, that's sort of baiting the poor ghost, but I suppose they didn't have anything better to do than torment the dead. You know, not like working their own fields or anything like that.

Folks who went near heard the spectre mutter the words, "What shall I do with this?" about the stone they saw he carried in his arms. One of the people who gathered there was a yokel, but apparently everybody else was too frightened to bother speaking to a spirit, so the man uttered, "Why, you damned fool, put it back where you got it!" Whereupon the ghost vanished and, we can imagine, there was much general rejoicing.

(Interestingly, a ghostly tale recounted in Dahlgren's *South Mountain Magic*, concerning tales in Frederick County, Maryland, describes the phantom of a farmer, who likewise moved markstones, which was laid to rest by the *exact same words*.)

The next tale is from Northampton County, and it concerns two farmers who were constantly at war over property lines. They moved each other's border markers, and swore that they even death would not stop their feud. The pair eventually died, and people on neighboring farms heard rattling chains and saw balls of fire flying along the borders between the two old farms. The flames crashed into each other with showers of sparks.

Both of these, however, were indicative of a much older tradition mentioned by Grimm in *Teutonic Mythology*:

> "[One class of ghosts were those] of men who in their lifetime dealt

wrongly by the cornfield, who respected not the sacredness of land-
marks. Unrighteous land-surveyors... may be seen hovering up and
down the furrows with a long fiery pole, as if re-measuring the wrongly
measured; whoso has ploughed of his neighbour's land, whoso has
moved the mark-stone, on him fell the curse of wandering as a will o'
wisp."

There was also a house in Limekiln (Berks County) that was haunted by ghost animals – a huge black cat without a tail and a nameless tan animal. Really, the only reason I mention that is that there was an old German superstition that limekilns were often haunted by spirits. And what could exemplify that superstition better than a *town* named Limekiln?

I also wonder whether the iron furnaces which dotted the Pennsylvania landscape were seen as a sort of "successor" to the superstitions associated with limekilns – and whether the body of ghostly lore associated with the iron-smelting industry is due to this succession.

Many older houses built by the Pennsylvania Dutch have a brick missing underneath the eaves. This creates an opening called a *seelenfenster* or "soul window" (see picture below). It was believed that when someone died in the house, their body would be placed in the room with the window, from whence their spirit could use the opening as a means to escape the

walls, thus preventing your house from becoming haunted.

This book would also be somewhat remiss if it did not mention the uniquely Pennsylvania Dutch tradition known as "powwow," references to which permeate the Susquehanna Valley and many items of folklore to be discussed in the pages to follow. Powwow, or *brauch*, is a system of folk healing most prominent in York and Berks Counties, but also prominent in Lancaster and Lebanon Counties. There is also a related tradition known as *hexerei* or hexing, the dark mirror of powwow. Powwow "doctors", or *brauchers* as they are known, are practitioners of a healing tradition nearly two centuries old.

In 1819, a German immigrant named J.G. Hohman wrote *The Long Lost Friend*, also known to more traditional Pennsylvania Dutch as *Der Lang Verbogen Freund*, the "bible" of powwow. The book combined Native American mysticism with German folk healing in its encyclopedic listing of remedies for nearly every possible health problem afflicting the early nineteenth century farmer or his livestock. *The Sixth and Seventh Books of Moses (Sechstes Und Siebentes Buch Mosis)* was another book that was an essential part of the *braucher*'s library, and *Egyptian Secrets* by Albertus Magnus another. Later powwow doctors also contributed their own books and pamphlets.

One copy of *The Long Lost Friend* played an integral role in the infamous "Hex Murder" at Rehmeyer's Hollow in York County which will be recounted in full in a later chapter.

Some of the more random and bizarre cures cited by Hohman and later powwowers are given below.

- If your cow refused to enter the stable, you were advised to bury a piece of wood, a handful of grain seeds, and a horse's skull in a box underneath the threshold – a needlessly elaborate cure for what amounts to a stubborn cow.
- To cure warts, one should take a whip, cut three notches in its handle, and bury it at a crossroads. The affliction would then be transferred to the person who happened to pick up the whip. This malefic act is not unknown in powwow; such transference of hurts is common. Northampton County powwower Peter Saylor (he referred to himself as a doctor, as did many powwowers, but this does not indicate any sort of medical training) as well as his student John H. Wilhelm claimed to have used the storied Hexenkopf (witch's head) Hill near Raubsville as a gigantic "receptor" or *zwischentrager*, both drawing power from it and banishing cured ailments into it. The diseases and ailments stored in the hill's rocks accounted for the sinister glow it possessed on moonlit nights (in actuality, mica deposits in the outcropping caused it to glitter and shine).
- If a horse was bewitched – several "magical diseases" or curses were said to afflict horses, among them such commonplace ailments as muscular atrophy or *swinney* and swelling around the leg joints or *windgalls* – the cure was so bizarre as to bear repeating here.

> "Take the bones of a dead person, from the cemetery, and a piece of wood that has been washed out by water. Then, take an earthen pot and pour into it a quart of vinegar and add a few scrapings of the bone and of the wood. Stir well with the wood and then pour into the horse's mouth, making him hold his head up so that he will swallow all of it. Bleed him at the shoulder vein. Tie some of the bone and of the wood on the right side under the mane. Return the rest of the bone and of the wood to the place where you got them."

Well, I suppose it's nice that when you advocate grave-robbing, you do at least advise the neophytic necromancer to put the body back where they found it, but that's probably small comfort to the poor person whose eternal slumber is disturbed!

- To cure a hernia, one should go to the county line. Straddling the line, the powwower should cut three hairs from the afflicted person's head, cut a small slash in a tree, and put the three hairs into the slash. When the mark in the tree grew shut, the hernia would heal.
- A cut on the foot could be cured by walking barefoot through a field and stepping in cow manure. To me, this sounds like nothing more than a recipe for infection.
- A curse laid on you could be lifted by taking a small wooden figure of the witch, driving a nail into it (but not through the heart) and burying it. The witch would gain an ailment of the area the nail was driven through and you would gain vengeance. However, first you had to know who bewitched you. The method of determining that, used in the days leading up to the Hex Murder (see later) was to glance at your hand after concentrating on a dollar bill over which the *braucher* had said an incantation. Supposedly, the face of the one who cursed you would appear on it
- Hex signs, that iconic symbol of Pennsylvania Dutch country and found so often on the side of barns or outbuildings, were originally called *hexafoos* or 'witch-feet.' As Wallace Nutting wrote,
  > "It is understood by those acquainted with witches that those ladies are particularly likely to harm cattle. As the wealth of the farmer was in his stock... the hexafoos was added to the decoration of the barn as a kind of spiritual or demonic lightning rod."

It seems that the hex sign, therefore, was meant as a sort of substitute or symbolic representation of the superstition circulating about toad's feet discussed in the last chapter.

German folklore and mythology was brought to these shores along with immigrants from that country, as I hope I have made clear. The traditions may have been given new flesh and recast to fit into the New World, but the core of the belief was old indeed. Later chapters will give more examples of transplanted folklore.

# Chapter Five
## The Bird of Happy Omen

In his *Sketches and Legends Pertaining to Bucks and Montgomery Counties*, William J. Buck related a tale he had heard as a boy in Montgomery County around 1835.

"It was stated that about midnight in winter with a tolerable snow on the ground, [a female relative] acted in the capacity of a nurse to the said aged woman, who was at the time lying very low and not expected to recover. A light was kept burning in the chamber on a stand near a window. Being otherwise alone and to beguile the wearisomeness of the time and as appropriate to so serious an occasion, she took up the family Bible and after reading therein probably for about half an hour she heard a peculiar noise at the window. Turning around to observe the cause, she observed distinctly a beautiful snow-white bird the size of a full grown domestic pigeon standing on the window sill nodding its head and gently tapping against the pane at least five or six times. She sat quietly without any fear, determined to observe it closely. After being there about two minutes it flew off and suddenly disappeared.

The good woman had often heard of several family traditions respecting the bird of happy omen coming to the sick chamber window even at midnight to announce a speedy and a happy death to the unfortunate sufferer, and this under the circumstances must be that messenger. On seeing the bird, as may well be expected, she somewhat excitedly went to the bedside to be now satisfied that her patient could not possibly long survive. She therefore hurriedly aroused the several members of the family, and related to them that she had seen the long-spoken-of white dove at the window, where he had left his token. Such an announcement, as may be well imagined, created some consternation, for the woman was actually dying and in less than two hours thereafter breathed her last.

This must have taken place at least over half a century ago, and could not fail under the circumstances to cause considerable comment in the family and among their numerous relatives. It was also the means of reviving quite a number of traditions of similar occurrences that had happened within a century previous, and all had prognosticated with an equal certainty, What else, they reasoned, could this be but a messenger from the spirit world?"

Buck then went on to describe how a neighbor of his had found a snowy owl in his hayloft, normally a resident of Canada. The owl "only came as far south as Pennsylvania in very severe winters for the sake of food, being abroad only at nights, in the day time remaining securely sheltered." To him, it was similar appearances of this far-flung bird that had "given rise to this beautiful if not romantic fiction."

While I'm not going deny the possibility that rarely-seen nocturnal birds like owls may have, indeed, been a part of the tradition of this bird (although I am forced to wonder whether the owl captured by Buck's neighbor could have been a barn owl rather than a snowy owl), another possibility presents itself.

Birds (owls and doves in particular) are commonly associated with death and dying in the folklore of Wales.

Doves circling above one's head were thought to portend death or illness and there were traditions of a variant of the banshee called the *aderyn y corff* (corpse-bird), which sits at the door and sings as an omen of death (this is similar to other prognosticators of death appearing at the window, such as the hag known as the *gwrach y rhibyn*; the *cyhyraeth* is another form of Welsh banshee). A very close parallel indeed is recorded by folklorist Marie Trevelyan:

> "The first-born daughters of an old Welsh race were turned into doves if they died unmarried, while the married ones became owls. The death of each female member of the family was presaged by the appearance of these birds or "by their bite" or pecking."

When the land now known as Pennsylvania was first colonised in 1680, one of the names that William Penn considered for the colony was New Wales; coincidentally, only four years later Penn was approached by John Roberts, a Welshman, who with Penn signed a document authorizing the creation of an independent county, which would speak only Welsh, rather than English. However, by the beginning of the next decade, the Welsh Tract was abandoned and the land divided into various counties.

Many places in suburban Philadelphia have names of Welsh localities. In Montgomery County, one finds such places as Bryn Mawr, Bala Cynwyd, Gwynedd, and Lower and Upper Merion Townships (named for the region of Merioneth); in Delaware County, Radnor Township, Berwyn and Colwyn; eastern Chester County also has some areas with names that, while not Welsh localities *per se*, are definitely derived from Welsh: Uwchlan Township (which, coincidentally, is the location of the stories of the Dorlan Devil), and Tredyffrin Township. The Welsh Mountains are located between Berks, Chester and Lancaster Counties. All of these cities and townships were within the Welsh Tract.

It should come as little shock that a similar superstition exists in the above-named Welsh Mountains; there, it is said that a white owl flying into and thumping against a window and sitting on the window-sill is an omen of death for someone within the house. As discussed above, owls are a common portent of death in Welsh lore.

It is indeed ironic that Buck, although a careful historian and apparently aware of the Welsh Tract, did not seize upon the status of the "bird of happy omen" as merely a New World version of a Welsh tale.

# Chapter Six
*Animal Superstitions*

- Farmers needed to slaughter pigs during a period of the waxing moon, otherwise the meat would shrink from the bone, wither and not be any good.
- Draping a dead blacksnake over a fence will bring rain.
- Cattle kneel at midnight on Christmas Eve.
- If a toad or swallow is killed on a farm, the cows will give bloody milk. This may be connected with another superstition that toads will suck the teats of a cow.
- A toad's foot nailed onto the door of a stable will prevent the entry of witches that may put a hex on the cattle.
- A witch will be barred from entry into your home if you feed the witch the ashes of the burnt ears of a black cat.
- If a "false tongue" falls out of a colt's mouth when it opens its mouth for the first time and it is hung in the stable, that colt will always be caught swiftly at night. This superstition seems to exist only in Fayette County.
- A cock's crow at any time other than morning signifies a male visitor, and two hens fighting signify a female visitor.
- A cat washing its face is another portent of visitors, or of the weather clearing.
- The death of someone in your house is prophesied by the chirping of a cricket within the walls.
- Another prophetic sign: horses playing in a field signify a funeral in the near future.
- A raptor, shot on the farm, with outstretched wings nailed on the gable of a barn keeps away other birds of prey. Similarly, in northern Pennsylvania, at least in the Sugar Valley of Clinton County, it is customary to nail a wolf's paw on the barnyard fence to keep away wolves.
- It was customary to shoot all birds of prey found on Ascension Day.
- Poison could be drawn from a snake-bite by tying a toad to the afflicted area.
- A spider hung from around the neck could cure ague.
- An eel skin wrapped around the limb would help a sprain heal more quickly.
- A child could be cured of bed-wetting if you could get him to eat fried mouse pie.

Unlike most other places, in the Welsh Mountains of Lancaster County, black cats were *good* luck; if one stayed in your house on its own accord, it was a good omen.

*Groundhog Day.* – February 2, the day of the Roman Catholic holiday of Candlemas, is the day the titular ground squirrel emerges from his burrow. If he sees his shadow, he retreats into his burrow, signifying six more weeks of winter; if he does not, he will emerge from the burrow, heralding the end of winter. Henry W. Shoemaker noted that these weather-foretelling powers were originally the province of the bear. This is now a nationwide tradition, but began as a Pennsylvania Dutch one.

# Chapter Seven
## Hex and Violence

In a previous chapter, I introduced you to the *brauch* (or powwow) that has subtly infiltrated and colored many aspects of life in the Pennsylvania Dutch region. Indeed, several tales of mysterious creatures in the lower Susquehanna valley may be connected with *brauch*. Some of the most infamous of these are connected with the Hex Murder, to which I alluded in that chapter as well. There were actually two Hex Murders, but I will begin with the most famous.

In 1923, John Blymire (often spelled Blymyer) of York County was judged as delusional. He had a deep-rooted belief in goblins and witchcraft, and he could not work, sleep, or do much of anything. He claimed to himself be a *braucher* whose magic was no longer effective due to his lifelong bouts with *opnema* ("take-off;" this was an ailment believed to have been caused by a curse, and one which likely is actually caused by malnutrition). Judge Fisher of the York County Courts sentenced Blymire to the state mental hospital in Harrisburg. After only 48 days, he escaped from the asylum (below), but the state apparently expended little effort to recapture him.

When Blymire returned to York County, he was not healed at all; in fact, it was reported that he would wander the streets looking for witches, and believed that evil spirits were crawling all over the walls of his dingy little apartment.

It was these delusions which led to his assault on his wife, who later divorced him; he consulted a series of *brauchers*, as he had before his stay at the mental hospital, in a vain attempt to learn the identity of his hexer, who it was who had cursed him.

In 1928, he met a boy named John Curry, and another young man named Wilbert Hess. Then he went to see the powwower who was to put into motion the wheels of his destiny. Emma Noll was an ancient Marietta woman who convinced Blymire that the witch who afflicted him with the wasting curse was a man named Nelson Rehmeyer, a supposed practitioner of the dark path of *hexerei*, and a man Noll said was also responsible for the Curry and Hess curses. To lift the curses, Blymire had to procure a lock of Rehmeyer's hair and his copy of *The Long Lost Friend*. He needed to bury the hair behind the Hess barn and burn the book.

Rehmeyer was a resident of the region of southern York County known, unsurprisingly, as Rehmeyer's Hollow (see below). He was a tall, gaunt old man, just the kind of fellow who looked like the "creepy old guy" in a horror flick. He was supposedly part Susquehannock Indian, and his family had inhabited the same cabin in a clearing in the woods for years. The Susquehannock connection is interesting – Eshleman, in his *Lancaster County Indians*, states that the Susquehannocks had been suspected by the other tribes of having created smallpox. He writes:

> "These mighty mysterious Susquehannocks were the frightful enemies of other tribes, and they very commonly attributed evils, misfortunes, and calamities to them, believing that the Susquehannocks had some

occult association with the devil and supernatural powers of many kinds."

One of his neighbors had told him of his magical pursuits, "Nelson, that crazy stuff is dangerous, and you ought to quit, because it will get you nothing but trouble." How right he was. I'm sure he didn't know just how much trouble, though.

On November 26, Blymire, Curry and Hess travelled down to Rehmeyer's Hollow, first stopping by to see Nelson's estranged wife. When she referred to him as a "devilish old witch", this doubtless strengthened the belief of Blymire that this was, indeed, the man he sought. For reasons that aren't exactly clear, the trio didn't procure the hair and the book the first time they went. The next day, they went back, this time armed with rope (Curry believed the plan was to take his hair while he was tied up).

On November 27, then, they returned to the cabin in the woods to procure these things. At some point, things went bad and turned to murder. As it turned out, the trio attempted to burn Rehmeyer's body – and the wooden cabin – but all was for naught as the fire extinguished itself. Physiological reasons for this were given, but of course among some it was Rehmeyer's status as a powwower that was the cause. And thus, the murder was discovered.

The resulting trials rocked York County and made national news. The headlines were none too kind to the tradition of powwowing. *HIGH PRIESTESS LEADS YORK'S VOODOO CULT*, one read. *YORK COUNTY HELD IN GRIP OF BLACK ART PRACTICES* was another. The headlines served to cast a pall over the entire proceedings; to ensure a fair trial was nowhere to be found. Setting aside any legal questions, however, all three of the killers were imprisoned. But during the trial, something bizarre came to light.

The ringleader of the little band, John Blymire, first mentioned it during his questioning by his attorney, Herbert H. Cohen:

> "When we got out to the porch, [Curry] says something about the key. We went out over the porch to the barn. When we got to the barn, we looked back and I said, "There's something standing in the road." They said it was a shadow of something. It started as though it backed out into the road. Then we all started to run as hard as we could up over the hill."

Blymire wasn't the only one. John Curry related the same incident to his attorney, Walter van Baman:

> "So we went out across the field, and when we got to the top of the hill Blymyer come to the conclusion that the house wasn't burning. He said, "Let's go back." So we went back, and when we got back to the barn, Blymyer and Hess thought they saw somebody in the road."

That's possible, I suppose, but we're left with the question of who would be standing in the road of a sparsely-populated neck of York County (at the time) like Rehmeyer's Hollow late at night. I'm certain that Blymire saw it as the demon freed from Rehmeyer's body, or some such metaphysical claptrap.

Except that it isn't an isolated event. In October, 1975, Lon Strickler was at the Rehmeyer house when he saw the dark shape of a man pacing back and forth across the road. After he called out to the shape, it ignored him at first, eventually looking up. As Lon realized with horror, the shape had no facial features on its head!

He said that he felt this was the ghost of Nelson Rehmeyer, and who is to say it wasn't, or that it wasn't the same spectral form seen by the three murderers that night in the winter of 1928? It is also interesting that throughout 1974 and 1975, Harford County, Maryland (adjacent to York County) was experiencing a period of increased Bigfoot activity.

The story of Rehmeyer's Hollow, or Hex Hollow as it is known to many people, has undergone a metamorphosis into a full-blown urban legend, and the stories vary wildly. Many people claim that the hairy humanoid monster known as Bigfoot roams the Hollow, leaving its stinky trails. That story may actually have some legs – although not from the Hollow itself, a triangular region of southeastern York County wedged between the Maryland state line and the Susquehanna River has produced a number of Bigfoot reports. These are likely the northern fringe of what seems to be a sizable colony or population group of the beasts existing in northern Maryland.

In October, 1971, Bob Chance (a prominent Maryland-based Bigfoot researcher) and seven others were near Castle Fin, along Muddy Run in southern York County, when several large boulders were thrown at them. Though the culprit in the rock-throwing incident was never seen, clues to its identity may have been given by events occurring seven years later in January of 1978, when a man living in Fawn Grove saw a 10' tall, stinking, hairy beast-man.

A month later, on February 2, Allen Hillsmeier discovered a trail of over 16-inch long, three-toed footprints on his farm near Delta. Upon further investigations, Hillsmeier found the partially-eaten remains of a rabbit and several dark hairs caught in a barbed-wire fence. Bob Chance also investigated this sighting and discovered more rabbit remains and more hairs.
Later Hillsmeier and a co-worker of his, Mike Asselin, founded a small organization called Bigfoot Investigations. They would find that a month after he discovered the tracks on his farm, a truck driver was approaching the Peach Bottom nuclear power plant when he saw "a whopper of a man" running across the road and disappearing into some trees. Guards at the power plant also heard a series of pig-like squeals and upon investigation turned up a number of three-toed tracks.

In December of 1979, some sort of animal tailed a hunter near the Holtwood Dam. The hunter saw a seven-foot tall creature crashing out of some brush, fleeing some gunshots that he had heard.

But back to Rehmeyer's Hollow.

Somewhere in the Hollow, many stories state, there exists a spot known as the Devil's Circle. It seems that somewhere in the forests of the Hollow is a circle of stones. The story circulating about this Satanic stomping ground (how's that for some alliteration?) is that if you go to them at midnight and walk in a clockwise circle around the stones, ghostly hounds appear. However, if you walk around the stones in a counter-clockwise circle, well, the hounds still appear, but this time they chase you down and rip out your heart.

This seems to be a bit similar to the story of the Witch's Circle in the Erie Cemetery. Then it also appears similar to the story of the Hans Graf Cemetery in Lancaster County.

Closely connected with the story of the Hollow, I think, is the tale of the Seven Gates of Hell. In fact, some folks locate the scene of the Rehmeyer murder nearby (though the Hollow is, in actuality, nowhere near most of the locations I've heard, and is what causes me to believe that this is a variant of the Rehmeyer story by someone who couldn't recall what actually happened. Maybe mixed up with another story, but more on that later). This supposed portal exists in York County, but the trouble is that nobody is really sure where it is. I don't know about you, but if there's a possibility of me buying a house and raising my kids next to the portal to the netherworld... I sort of think that's the kind of thing I'd like to know *before* I buy. Most people will tell you that it's on Toad Road. The only problem is that there is no Toad Road in York County. To paraphrase Joe Walsh, it's hard to go to Hell when you can't find the door.

Well, the typical story goes like this: you have an old mental asylum in the woods. Oh yes, those old nut hatches in the woods, there's a ton of them... and like every good old woodland asylum, it burnt down. Several of the inmates escaped and ran rampant through the woods, before they were killed by the very folks that came to rescue them. Nice. Remind me not to call for help in York County anytime soon. All the deaths called up some nasty spirits. Though some will say that there was a house where the mother killed her husband and kid before hanging herself. So was this in addition to the asylum, or instead of it? I'm not sure.

In any case, it's usually said that the town fathers of York placed six gates along the path to the old asylum to discourage anyone else from ever going there. What? But I wanted to go to the ruins of that old burnt-out loonie bin, dammit! As you might have guessed, these are the Seven Gates of Hell, the seventh being the threshold of the old asylum. I thought it burnt down? Or do they mean the murder house? The Seven Gates of Confusion and Obfuscation, is more like it.

Supposedly, each of the Gates cause something eerily supernatural to happen. You know, like weird lights and ghosts that tell you to turn back, and lots of really wild things. It's supposedly an eight-mile hike from the first Gate to the old asylum.

I took a trip there once, using one set of directions I got off the Internet. It put me on Trout Run Road, near a rusted cattle gate nearly hidden by weeds. This, I was to find later, was usually said to be the first Gate. Really? But the story could have some legs and still more cryptozoological relevance, and the directions weren't even that far off.

The end of Trout Run Road does a 90° turn into Range Road – *now*. But on 1700s maps of Hellam Township, the road used to continue northward, along Codorus Creek, to intersect what is now Furnace Road (leading to Codorus Furnace, a Revolution-era iron furnace). Indeed, some of the versions of the legend say that people were hanged in the woods near an old furnace. At the intersection with Trout Run Road, there used to be an old building. That building was still there when I was younger, though sometime in the early 1990s it burnt down. That building was far too small to be an asylum. I searched my family's collections for years for the photograph we had of the building, and finally it turned up in 2010.

My grandfather recalled how in the 1950s, the building had been lived in by an old man who said he was once a doctor. A hermit. People used to bother him (as they are wont to do with hermits). My conclusion is that the story of the Seven Gates is a conflation of the Rehmeyer's Hollow tale and a distorted memory of driving back the road to harass the hermit.

Like Rehmeyer's Hollow, the Gates are said to be the stalking grounds of Bigfoot. Indeed, one of the eerie occurrences noted from the Gates area are unearthly screams, which could originate with such a creature. Sometime in the 1970s, Bigfoot sightings took place near an old furnace, and in a forested area known as Condoor woods (Kondor being the family who owns much of the woodland in the Codorus Furnace area and along the former extension of Trout Run I mentioned, which may have been the fabled 'Toad Road'). Whether this is a valid sighting or another example of confusion with the Rehmeyer story I'm not sure on.

A grayish form with glowing white eyes – Bigfoot or apparition? – was seen around the Seven Gates (though I'm not sure where was meant by the witness, the false gate on Trout Run or the inaccessible road) in 1999.

In 1934, Albert Yashinsky of Shenandoah shot and killed Susan Mummey ('Old Suss') of Ferndale seven years after she supposedly hexed him as he worked in a field adjacent to the Mummey farm. Yashinsky is said to have muttered in prison, "Those eyes! Oh, those eyes, oh how I wanted to have them closed! I could not stand them!" This incident was also connected to an appearance of some weird zooform creature, though in this case it is more obvious.

Yashinsky claimed that 'Old Suss' had burdened him with a hex in the form of a monstrous black cat with huge green eyes and the witch's face. For seven years, he had been climbing into his bedroom through the window, as otherwise "unless he trod evenly on each step, the cat sprang out at him." He also claimed that at least once per month, the black cat visited him as he slept:

> "It slowly crawled through his closed bedroom window and towards his bed. Then it would rest itself on the side of his bed and claw at his side. It was painful torture... Once a month and sometimes more often, this huge black cat would visit him and make it impossible for him to sleep... after a visit from this cat he would be completely lost and bewildered. He was actually helpless and unable to work."

Apparently, the murder of 'Old Suss' was not effective in the banishing of this cat. While in prison, Yashinsky claimed to still be visited by this fiery-eyed, witch-faced phantom, rather like a feline equivalent of Keziah Mason's infernal familiar, Brown Jenkin, in Lovecraft's *Dreams in the Witch-House*.

These black felines have a lengthy pedigree in the annals of *brauch* and *hexerei*, thanks to an overgrown foundation in the Tumbling Run Valley, near Mount Carbon, again in Schuylkill County. These are the remains of the Thomas Farm, and in September of 1911, with the death of Pottsville native Howell Thomas, another story of witchcraft and hex came to light.

Although Thomas' death was attributed to a stroke, his daughter, Mary Isabella, attributed it to witchcraft. She had received a warning from an elderly practitioner of powwow that she could expect a visit from a black cat, one sent by "an enemy from Orwigsburg on a mission of evil" and, as a result, she had taken to carrying a revolver on her person at all times. When it finally appeared, it hissed at her and Mary Isabella shot it. But the bullet failed to have the expected effect, causing the cat to swell up to five feet long. The witch-cat then did its evil work. The cows sickened and died, followed by the horses, apples, and vegetables, and finally culminating with the death of Howell. At his funeral, Mary Isabella accused one of her sisters of being the Hex Cat (or the witch behind it, this part of the story isn't clear), and apparently, Howell felt the same during his lifetime. Another powwower told Mary that the Hex Cat could be killed by a golden bullet, but after this was procured the bewitched cat failed to appear.

At any rate, Mary and the other hunters killed a black cat, which was nailed to a barn. This was likely the black cat owned by a Schuylkill Haven man which was called by powwowers a 'hexahemeron cat'. The cat was supposedly born on June 6, 1906 (6-6-6) as part of a litter of six kittens. The Thomas Farm was called the Hex Cat Farm in popular lore, and countless trips by locals were taken to the tumbling remains of the farm to see the decomposing remains of the black cat. A rather morbid pastime, to be sure.

Another member of the clan – William S. Thomas of Pottsville (landlord of Howell Thomas' farm), was also burdened by the Hex Cat affair. It seems that after Howell's death in 1911, William felt that the Hex Cat bore his family a vendetta. The bewitched feline haunted William and five years later until he attempted arson in an attempt to dispel the cat. William was convicted in 1916 of attempting to burn down a number of properties he owned near Third and Race Streets in Pottsville. He was sentenced to an unbelievable *three months* in prison, a punishment surely unbefitting the crime! After he was released from prison, he became a hermit, and took up residence in a shack near the Tumbling Run resort. Here, he was found frozen to death in January, 1918.

It has been well-established that Berks County is one of the regions of Pennsylvania in which *hexerei* or powwow is the strongest. In fact, *The Long Lost Friend*, the "Bible" of powwow, was first published in Reading. Several known powwowers have lived in the county, among them "Mountain Mary", Anna Maria Jung, one of the most famous and powerful *brauchers*, who lived in the Oley Hills.

On November 3, 2009 Mindi and I took a trip to Reading, in Berks County, to get a bit of Christmas shopping done. A few days previous, I had found out the location of the supposedly haunted Union Canal in Reading. I was probably seven or eight the first time I heard the story of the canal and its ghosts on a TV special I was watching about local ghost stories. As the canal wasn't all that far from where we were going to be, I suggested we make a stop so that I could check out the area. I'm glad that I did.

Louise Bissinger, wife of prominent Reading brewery owner Captain Philip Bissinger, was already distraught. Her husband had been unkind to her for years, and was openly conducting an extramarital affair he wasn't even attempting to hide. The two children the couple lost (in 1869 and then again in 1874) placed even more stress on her, and it is likely a combination of these three things, as well as the fact Louise was pregnant with another child, which led her to take the actions she did one fateful Tuesday. On August 17, 1875 Louise Bissinger took her three living children (ages nine, six, and three) on a walk. It was to be their last.

They crossed the Schuylkill River at the Penn Street Bridge, walked along the river, then turned northeast along the Tulpehocken Creek. When they reached Lock 49 of the canal,

Louise filled a basket with rocks and tied it around her waist, grabbed her children, and plunged into the canal. The canal was shallow, only seven feet deep, but the sole witness could not swim. By the time a boat could be launched from the home of miller Samuel Gring, Mrs Bissinger and her three children were dead.

Later, Mr. Bissinger – who looked rather Luciferan in his picture from the newspaper articles reporting the tragedy – was said to have mourned the deaths of his children, but not his wife. He later remarried, and his second wife Ida is buried with him. Louise is buried with the children. The 1875 tragedy may be partly what spurred him to will a fairly sizable sum to various orphanages when he died in the 1920s.

It should come as no surprise that such a tragic occurrence would spawn ghostly legends, and Charles J. Adams III (a journalist for the *Reading Eagle* and author of several books on the ghostly lore of various regions across Pennsylvania) has written extensively on the various ghost stories of Berks County. Typical occurrences at the canal include the feeling of being followed along the towpath, presumably by the children whose apparitions are sometimes seen there.

The Wertz Mill Bridge, also known as the Red Bridge, leads to a cluster of old homesteads and then to the Union Canal towpath. The wooden covered bridge spanning the Tulpehocken is actually one of several Pennsylvania "crybaby bridges," a familiar trope of urban legends throughout the East Coast and, presumably, worldwide. Legend has it that a single mother tossed her children over the side of the bridge – a bit hard, since it's a covered bridge! – and that sometimes, pedestrians can hear the murmuring cries of the children. The "cries" are probably just the sounds made by flowing water, and as the bridge is less than a mile from the scene of the Bissinger drownings, it's quite possibly a distorted version of that story started by someone who couldn't quite remember what really happened that day in 1875. The trees flanked the towpath and created a tunnel effect. It seemed to be perpetually dusk on the path – now a paved trail traversed by joggers and bicyclists alike. The trail was only a matter of 10 or 15 feet from the creekside, but even that was barely visible at times. The canal itself had been completely drained. Indulging my curiosity, I naturally veered off the path into the bed of the canal itself. Of course, I wasn't really anywhere near the haunted lock yet, and in hindsight, I wasn't going to walk nearly a mile through brambles and dense undergrowth, so after taking a few pictures near what was once Lock 50, I got back on the towpath.

The whole way towards Lock 49, I heard sounds in the brush at the edge of the canal intermittently, and had a feeling of being followed. In my mind, I made this into a ghostly encounter, since the ghosts of the Bissinger children are supposed to follow pedestrians, but as I later found that they approached the lock from the opposite direction, I'm fairly certain that these noises were probably squirrels invisible to me and that my mind had done the rest.

What justifies the inclusion of this ghostly travelogue in a book about cryptozoology? Nick Redfern had written the book *Man-Monkey* about his investigations into a harrowing encoun-

ter which took place on the Shropshire Union Canal. It seems a nocturnal traveler had been severely frightened by an ape-like creature which leapt onto the back of his horse while he was driving a wagon over a bridge on January 21, 1879. It was later conjectured that the "ape" was in actuality the spirit of a man found drowned in the waters of the canal. A month or two previous to my trip, I had mentioned to Nick the coincidence of *another* Union Canal here in the States, also the site of a suicide, and also with ghostly legends of its own. (That book revealed that the English story was to become even more eerie as Nick turned up a story concerning a suicide followed by the appearances of a shaggy man – also in 1879, oddly – at the railroad bridge over the Firth of Tay in Scotland, later the scene of mass death as the bridge collapsed and the train travelling over it fell into the North Sea.)

As I mentioned, earlier I had stopped in Lock 50 and taken a few photographs. What I didn't mention then was that one of them was of a number of interwoven sticks similar to formations often found at Bigfoot sighting areas and presumably made by that anthropoid ape. True, it wasn't nearly as impressive as some of those structures – but to me, at least, these sticks didn't look as if they had just naturally fallen this way, either. Needless to say, after seeing this, the parallels to the case of the English Man-Monkey took another turn. *Two* Union Canals, both with suicides, and both possibly connected with shambling sasquatch? This was beyond coincidence.

As I was soon to discover, the parallels were even more eerie than I previously had thought.

One of the creepiest sections of Nick's book focused on "The Cult of the Moon-Beast," a clandestine cabal summoning up all manner of monsters to carry out an agenda of assassination. It turns out that just such a cabal also existed among Reading's *brauchers* in the early 1940s. Well, I don't know about their summoning monsters, but curse-laying and assassination, yes!

Dr. Albert Mattern was a practicing physician in Reading in 1940-1942. He published an *exposé* of a group of elderly women which he called "The Committee." These aged *brauchers* had unusual beliefs:

> "The people who were part of the Committee thought that if they would put a hex on somebody they would have eternal life. A man asked me how to get rid of water running under his skin. I treated him for alcoholic neuritis, and it went away... [later] he found out what caused it. He found his mother-in-law running water on his picture, and that's what caused the feeling in his arms... At the same time I was treating his wife, and would give her an envelope with pills... She asked me to change to plain envelopes without my name or address, because if her mother finds out where she's going, she'll put a hex on the medicine, and it won't do any good. I was about the fourteenth doctor she had been to."

Dr. Mattern described their motives in another passage: "The Committee believed that as long as they could push another old person over the rim, they would never die, they would have life eternal." They were vampires of a sort, life at the expense of another.

Another example of a Berks County hex was a local urban legend in the Oley Valley. It was said that, if one would shoot a forest beast (apparently, any sort of animal would do) with a silver bullet, the Devil himself would appear. A bunch of young guys shot some unfortunate creature and a black-clad man with goat's feet and horns upon his head appeared in a road nearby! The young men, though, had the gall to summon Satan himself, but ran away to leave Old Nick standing alone in the road. Who knows what sort of "deviltry" he got himself into!

During World War I, this legendary hex was altered in true urban legend fashion – except this time, rather than summoning the Devil, it would cause the death of Kaiser Wilhelm. Or maybe it would just compel him to assassinate the German (I'd think that not only summoning, but actively exploiting the Devil himself would require more than a bit of animal cruelty, but that's me). Either the hex was a crock or Beelzebub's not that great a murderer, since Kaiser Wilhelm didn't die.

The documented sightings of the hairy guy in Lebanon County – sightings in Lebanon, as well as Annville – were all from along the Union Canal's former pathway. It has been well-documented that many species will utilize creek beds and railways, among other natural and manmade pathways, to mark their passage through the countryside. Could Bigfoot be utilizing the canal bed, overgrown though it is, as a pathway?

# Chapter Eight
## Bizarre Tales of the Animal World

In the last few years, there have been a few reputedly monstrous hogs killed in the Deep South. Most of these giant pigs have been hybrids of wild and domestic swine. First there was the so-called 'Hogzilla,' (supposedly) a gigantic 1,000 pound beast killed near Alapaha, Georgia in 2004. Recent examination of the hog's remains revealed that it wasn't quite as big as was initially reported – probably somewhere in the vicinity of 800 pounds. That still made Hogzilla quite a hefty specimen, however.

Another, and one that is coming under legal scrutiny, is the cleverly-named 'Pigzilla,' a boar supposedly 9'4" in length, and 1,051 pounds killed by an 11-year old boy in Anniston, Alabama in 2007. It was quickly pointed out that the photograph, however, utilized forced perspective to make the hog seem larger than it really was. The boy in the photo is actually several meters behind the pig. Once it was determined that the pig in question was actually a domestic animal named Fred, the people in the case came under scrutiny for suspected animal cruelty.

Then there was another Georgia boar weighing in at 1,100 pounds, and then a Florida one weighing in at 1,140 pounds. Each tried to outdo the last. And though pigs are known to reach gigantic proportions, most of the largest ones are domestic, where they can be better-fed in someone's farmyard. A few regimens of growth hormones don't hurt, either.

That said, boar-hunting has been a pastime of kings for time immemorial, and tales of monstrous swine have been told, retold, and passed down for centuries. From the monster boar Twrch Trwyth, hunted by none less than King Arthur himself, to the Beast of Dean, a boar which haunted the region around Parkend, England and was reputedly so large it could knock down trees and overrun fences, to the hogs reputed to inhabit the London sewers, the British Isles in particular have been overrun with folkloric porkers.

Huge pigs seem to have been a popular topic of news coverage throughout the 1800s and into the early 1900s. I suppose that should come as little surprise in an agricultural area like Pennsylvania. Many of the pigs were large, but not outsized; however, there were a few monsters rivaling – and exceeding – the gigantic hogs of today. Some of the larger ones described in the news of the day are listed below. Nearly all of these surfaced from Lancaster County. They are given in ascending order of weight.

- A 501-pound pig was reported by Joseph Weaver of Allentown (Lehigh County), in 1837.
- Henry Newswanger, from Union Grove raised a 523-pound hog in 1920.
- In Goodville, also in 1920, Frank G. Martin reported that a hog reached a weight of 710 pounds.
- Michael and J. Groff had a hog that reached more than 800 pounds.
- A pig raised by Jacob Mellinger of Strasburg was 847 pounds.
- A 936-pounder was killed on the Marks Farm in Palmyra (Lebanon County).
- David K. Weaver, from Earl Township, reported a pig weighing 1,220 pounds.
- Finally, the largest swine was described in 1859. It was reported that a Mr. Bishop, of Lancaster, had a hog weighing an incredible 1,635 pounds!

In 1915, one of the melodramatic stories which frequently appeared in the news of the day appeared, describing a hunt for a pig gone feral. The pig had escaped from a farmyard the year before, and took up residence in the forests and marshes around Village Green (Delaware County). It had quickly become a wild boar, bedecked with huge tusks. It was finally killed by Jesse Raynor, and weighed nearly 400 lbs. Small fry compared to some described earlier.

In 2001, Howard Smith of Conoy Township (Lancaster County) shot a massive hog, a cross between a feral pig and a Russian boar, which had been kept domestically. As it grew older, it became more aggressive, and when it escaped one day Smith decided to shoot the hog. It weighed upwards of 1,100 pounds and was 10 feet long.

All of which brings us to discussion of a very real problem in Pennsylvania, that of feral hogs. Around 2002, a number of domestic pig/Russian boar crosses (much like the one described above) escaped from a game preserve in Cambria County, and there also was a preserve in Susquehanna County where hogs escaped. The hogs established themselves in the wild (the Pennsylvania Game Commission reports that they are present in 18 of the state's counties, and at least two have breeding populations).

Black bears are the only species of bear native to Pennsylvania, although Henry W. Shoemaker notes that white-colored specimens thereof were killed in both Snyder and Susquehanna Counties in 1802 (I should stress that these are not described as anything other than black bears of unusual color) and that a hunter named Benjamin Lowe killed a white-faced black bear near Woolrich (Clinton County). He also makes reference to what was called a *fox-bear* or *red bear* but which was merely an erratically-colored specimen of black bear. One was

**OPPOSITE: This massive feral hog was killed in 2001 in Lancaster County.**

killed along White Deer Creek (the creek runs through Union and Clinton Counties, and he doesn't mention exactly where along the waterway this was), date unknown. He goes on:

> "The last red bear, and perhaps the finest of all time, was shot by Edgar Austin Schwenk, of Eastville, Clinton County, on the old Buffalo Path, Union County, November 29, 1912. The animal... resembled a fine Canadian Red Fox [and was] various shades of lemon, tan, and fulvous."

Another type mentioned was called a *dog-bear*. Jesse Logan (1809-1916) shot one of these on the Cornplanter Reservation, a Seneca Indian reservation lying between Warren and McKean Counties in the northwest part of the state, inundated during the construction of the Kinzua Dam in 1961, which now lies beneath the waters of Lake Kinzua. Another undated incident, Shoemaker notes that the dog-bear was "more warlike than the shyer and more inoffensive hog bear, and more akin to the semi-mythical naked bear." This may be the same as what is referred to as a *ranger bear* in New England. The ranger (racer) bear appears to be a genetic variant of black bear.

Of course, there are always bears found in places where they're just not supposed to be. Like along Centerville Road in the suburbs of Lancaster (Lancaster County), sometime in the early 2000s. These usually turn out to be young bears who've been separated from their mother through some mishap – perhaps a hunter killed the mother, or just by chance the cubs wandered too far away – and in search of the easy meals provided by human habitation.

The last bear in the Blue Mountains near Harrisburg was killed in 1910, according to Shoemaker; but try telling that to the virtual torrent of bears which roamed through the city's suburbs a century later. The first of the bears was seen wandering through residential areas in Upper Allen Township (Cumberland County) on May 8, 2010. That bear was struck by a car the same day, and wildlife officials feared it might have been killed before they managed to track it to a wooded area near a cemetery and capture it. Following on the heels of that incident, May 23 saw a black bear ambling along North Mill Street, just outside Annville (Lebanon County). The very next day, a wayward ursine was seen in Hampden Township (Cumberland County) and then on May 26 yet another bear was captured from a tree in Memorial Park in Lemoyne, same county. The bears took nearly two weeks off before yet another was captured right behind the Boscov's department store at a shopping mall in Camp Hill, same county.

The coyote's presence in Pennsylvania is not controversial. What is controversial is *when* exactly they appeared here. The official stance of the Pennsylvania Game Commission is that the coyote migrated southward from the Catskill Mountains of New York, and that the first verifiable kill of a coyote in Pennsylvania was in Tioga County in 1940. In *Extinct Pennsylvania Animals, Volume I*, Shoemaker reproduces a photograph of one killed in Clinton County by Amasa Winchester on December 16, 1915, and testimony by C.W. Dickinson, a wolf hunter from McKean County, in the same volume, indicates they may go back even farther

**OPPOSITE: The coyote killed in 1915 by Amasa Winchester.**

than that. According to Dickinson, a fellow from Bradford in the 1890s had a cage of five coyotes received from a friend in the Western states. They escaped, and four were later shot by hunters, although one was unaccounted for. (The coyote killed by Mr. Winchester, as it happened, was an escapee from Shoemaker's estate near McElhattan.)

The presence in Pennsylvania of three separate kinds of wolf, in the minds of settlers at least (these differed only in color of pelage, not anatomically) was established by Shoemaker; as it is noted that the "little brown wolf" barked, rather than howled as other wolves, he speculated as to whether this was actually a reference to an earlier population of coyotes.

Wolves may have been extinct in the state for over a century, but as of 1940, when the Federal Writer's Project guidebook to Pennsylvania was released, there was a highly-populated wolf sanctuary near Mt. Jewett (McKean County), the Lobo Wolf Pens. Currently, there is a sanctuary in the mountains between Lancaster and Lebanon Counties. It's possible that vocalizations from these wolves could, to some extent, fuel the story of the ghostly hounds of nearby Cornwall Furnace.

A species which doesn't seem interesting on the surface, the otter was formerly restricted to the Pocono Mountain region, but as of July, 2009, the aquatic mustelid's range was widening. Michael Reeder, a York County game warden, had found two dead otters along roadsides near Dillsburg. "I have been hearing of more and more otter sightings in the area and this shows we have an increasing population," Reeder said.

The wolverine is a species formerly native to the state, though recently extinct. They were always restricted to the northern counties and it has been listed among Pennsylvania's former species for only a century and a half. Potter County hunter LeRoy Lyman captured several wolverines, and C.C. Burdette killed several along the Pine Creek. Hunter Mike Long also supposedly killed several in Tioga County. The last one was killed in 1863 along the Sinnemahoning Creek's First Fork by Joe Nelson.

However, a friend of my grandfather's says that he saw an animal, which he supposed was either a wolverine or a badger, sometime in the 1990s moving up a mountainside in Potter County. Whichever it was, it would have been significant, as the badger is also of uncertain provenance in the state – and the last records of it, in any case, were from the southwestern corner of the state near West Virginia. The wolverine identity is indeed tempting, given that the last wolverines in the state were reported from this selfsame area. The deer population in the Northern Tier, as that region of the state is called, is almost definitely high enough to support a small population of wolverines, though likely not a large one.

In his "Remarks and General Observations on Mercer County, Pennsylvania" (1850), B. Stokely, who claimed to be the first permanent settler in the county, wrote that "an animal called the wolverene (*sic*) [was] supposed to be engendered between a fox and a wild cat".

**OPPOSITE: Jesse Logan, the aged hunter who killed a "dog-bear".**

A binturong was found underneath a porch in Beaver County in 2002.

The works of Robert R. Lyman are very much a "Ripley's Believe It or Not" of northern Pennsylvania. Some of the accounts are doubtless from the realm of folklore, but all are interesting.

In 1818, a huge herd of bison (still inhabitants of Pennsylvania in those days) trampled through Emporium (Cameron County), flattening outbuildings and destroying unfortunate livestock caught underfoot. The herd of bison eventually ran through a home, the cabin of Samuel McClellan. His wife and three children were trampled to death before McClellan and a posse of hunters killed the marauding herd.

Almeron Lyman of Potter County told the story of a smoky-gray fox that residents called the Dusk Fox. He tracked the Dusk Fox for years, eventually setting traps for it in 1904. It evaded nearly every trap he set for it, and when it finally was trapped, it bit off four of his toes and was never seen again. This color is a recognized morph of both the red and Arctic fox.

George F. Hart tells the story of how in 1888, his logging crew rescued an injured crow, which they named Jim, from a felled tree in Potter County. As George wrote in 1956:

> "...[Jim] soon learned to call me by my full name and would say, 'Hello, here comes George Hart. I know you George Hart'...it was not the words alone which he spoke so plainly that was a puzzle to us, but his actions. When he disliked anyone, he simply told them, 'I don't know you' and that settled it. There were men in my crew that he liked and a few that he disliked, with whom he never made friends."

On one occasion, Hart said, Jim yelled "I don't know you" at a wild crow which came too close and the second crow obediently flew away. On another occasion, he called for help when he was attacked by an owl.

> "Through the week, the cook would rap on the stove pipe and call: 'Turn out' when he had breakfast ready. On Sunday morning Jim would rap on the window and call 'Turn out' when he thought we were sleeping too late...One day the cook tried to put him out but Jim gave him a merry race so he put the dog after him. He got him as far as the kitchen where we had a row of wooden barrels containing groceries lined up along the wall. When the dog got too close, Jim dodged in between the barrels through the space made by the bulge of the barrels and where the dog could not enter. He then stuck his head out and said: 'Hello, I don't know you' and laughed as hard as he could scream. He could bark like a dog and his laugh was exactly like a man with a good, strong voice."

Another talking animal was Tucker, a starling kept by Joe Galloway. In 1933, Galloway and the bird lived at the Eden Nursery in Port Allegany (McKean County). Tucker didn't seem quite as conversant as Jim: the only thing he could say was his name. But like Jim, he wanted nothing to do with wild birds of his type.

Mynas are one of the only species of bird which can "talk" (really, repeat sounds), and are a species of passerine bird related closely to the starling, so the account of Tucker isn't at all outside the realm of possibility. In fact, the common myna (*Acridotheres tristis*) was, indeed, introduced to North America in the late 1800s, although their populations are limited to Florida (and the Hawaiian Islands). Crows, though not directly related to starlings, are still passerines and have been noted for their intelligence. Crows have been known to speak simple words, though whether one could speak to the extent of Jim is up for debate.

Another weird story of the corvid kind concerns two ravens at the Coudersport (Potter County) Golf Club, who took between 20 and 30 golf balls from patrons. After the groundskeeper shot one of the ravens, the second flew off and wasn't seen again.

One tradition which is long-standing is the idea that animals will feed their kinsmen which have been captured by hunters. In 1915, E.R. Duell and Charles Clark found a trap in Sweden Township (Potter County), set by a hunter the year before, containing a live fox which had been sustained throughout the year by food brought to it by its fellow foxes. The others had brought it two rabbits, a groundhog and several mice. However, this is a well-established folktale.

A binturong is just about the last animal most anyone would expect to find hiding in their backyard just waiting to be discovered on an early-morning mail retrieval. Unlikely as that seems, this is exactly what happened in Economy (Beaver County) in April, 2002. Joan States went out to get the newspaper one morning when her dog began sniffing at an unusual animal that emerged from underneath a bench near her barn. The creature was "mostly black with a white spot on its head, white whiskers and large paws". A call to the local police yielded a visit to the States' home by officers of the Pennsylvania Game Commission, who captured a four-foot male binturong (*Arctictis binturong*), also known as a bearcat.

The binturong, named "Binny," had escaped from the home of Brad Wilfong in New Sewickley. He said the animal had been acquired from a petting zoo in the Chestnut Ridge area in 1999. It is illegal to own a binturong in Pennsylvania without special permits. After his capture, the bearcat was kept at a pet store in McKees Rocks (Allegheny County), and was later moved to the Pittsburgh Zoo & Aquarium. Then he was moved to a facility in Dillsburg (York County) for two weeks and then it was back to the Pittsburgh Zoo before ending up in Austin, Texas.

Binturongs are a type of civet cat, which despite the name are not felines at all. Civets are believed by some to have been the origin of the SARS virus which swept through eastern Asia in 2003 (though most scientists feel the virus actually originated with another species).

# Chapter Nine
*The Broad Top Snake*

Pennsylvania is home to a few slithery creatures of the reptilian kind. Some of these are 'mundane' giant snakes, while some of those snakes may be real encounters with large snakes upon which have been placed a template of the dragon-slaying myths of the Germanic homelands of many Pennsylvania residents. And one seems to be of a genuine, St. George-style dragon, although even that may not be what it seems on first blush.

First, a bit of clarification is necessary: as Chad Arment states in the introduction to *Boss Snakes*, the term 'blacksnake' is a rather generic one which in everyday use referred to any one of a number of species of snake. In practice, any dark-colored snake was called a blacksnake, regardless of what it actually was. The term is most often used to describe eastern ratsnakes (*Pantherophis alleghaniensis*), coincidentally allied to the Aesculapian snake (*Elaphe longissima*) occasionally proposed as the 'true' identity of Gunnar Olaf Hyltén-Cavallius' *lindorms*. Ratsnakes can reach lengths of nearly 9 feet, making even some 'giant' snakes not much of a giant at all.

The most famous of Pennsylvania's giant snakes lairs in southern Huntingdon and northern Bedford Counties, and is the one from which this chapter takes its name. The 'Broad Top Snake,' as it is called, has been reported consistently for over a century and a half. The first sighting was in the 1830s, when a hunter in Cumberland Valley Township (Bedford County) met with a snake between 20 and 30 feet long "with head erect and [moving] at great speed."

Since that time, both Arment and Bedford County newspaperman Jon Baughman have logged twenty separate sightings of giant snakes in the Broad Top area, most recently in 2002 off Enid Mountain Road (Bedford County). Most of the sightings of the Broad Top Snake seem to be located in a roughly triangular area between the towns of Coaldale and Saxton (Bedford County) and Coalmont (Huntingdon County). Of the logged sightings, 2 give the creature's size at between 10 and 19 feet; 8, between 20 and 30 feet; and 2, over 30 feet; and 8 neglect to mention size at all or give an unspecific size. When coloration is mentioned, which it rarely is, the serpent seems to be black or dark gray in color, with yellow markings around the face (another report mentions 'white markings' but not where on the snake these were located). Another sighting recalls its color at a 'pale tan.'

Assuming a bit of exaggeration, we are left with a 15-20 foot snake. Arment feels that the Broad Top Snake is (assuming it's real, of course) of the genus *Pituophis*, which is comprised of bullsnakes and pinesnakes. In fact, a dark gray with yellow or white markings coloration would correlate remarkably well with the northern pinesnake, *Pituophis melanoleucus melanoleucus*. These snakes are quite large, with the record-holder being 8 feet, 9 inches and rumors of specimens over 9 feet in length. The Broad Top Snake would certainly be very large even for these large snakes! And though *Pituophis* does not currently occur in Pennsylvania, a fossil specimen is known from – guess where? – Bedford County.

I should here mention the Huntingdon Standing Stone, if only because the removal of such stones has been connected with all manner of monstrous appearances in other countries. The stone formerly stood near the city of Huntingdon, a fair distance north of most of the Broad Top sightings. It was still standing in the days of Conrad Weiser's journey to western Pennsylvania (about 1750). A stone now stands in the center of town, but it is a replica of the original. A similar monolith stands on the Oneida reservation in New York. It's not known for certain if this is the Huntingdon Stone or not

> "[Weiser] crossed the Juniata River, and approached the 'Standing Stone.' This was a prominent landmark of the region, and stood on the right bank of a creek of the same name, near the present town of Huntingdon. It was about 14 feet high, and six inches square, and served as a kind of Indian guidepost for the region."

Another location in which many huge snakes have been seen is southern Pennsylvania, just above the Maryland state line, reaching through the hills between Adams and Franklin Counties and into York County. The earliest of these comes from the region south of Gettysburg (Adams County), as related by Emanuel Bushman to the *Baltimore Sun* in response to sightings of another monster snake at Hall's Springs, Maryland:

> "One sunny day in April of 1833 my brother, with six others, were exploring Round Top [I believe this is Big Round Top] for the first time. The hill and surroundings were covered with a dense forest. As they were ascending the west side they suddenly came upon a monster snake, sunning itself upon the rocks. Part of them took to flight, but brother and two others stood to see how it would end. They described it as a black snake, apparently turning gray from age... They estimated its length to be from fifteen to twenty feet, and the thickness of an ordinary man's waist. They threw at it from above and it rolled down into its den. Father saw it about a mile from there at the big rocks, called the Devil's Den. Frank Armstrong saw it and was badly frightened. Grandfather saw it in his time, and mother says, the Indians used to speak of it as 'heap big snake.' Mr. Michael Fry, living near Round Top, saw it about thirty years ago [before the publication of the letter, or about 1845], which is the last time I heard of it."

Of course, Gettysburg became famous (or is that infamous?) about ten years before Bushman's letter saw print as the site of the terrible Civil War battle. This letter seems to suggest that the cluster of boulders known as Devil's Den, (above) used as a hideout by scores of Confederate snipers on the second day of the battle, had this infernal moniker before the events of that bloody July day in 1862. Another account by Salome M. Stewart suggests the same; that 'the Devil' was a massive snake which had his den among these boulders. It is thought that a battle was fought in this same region by Native Americans long before the town of Gettysburg was founded, and there is certainly no shortage of ghostly tales born from this carnage.

But moving on from this momentary diversion, Solomon Smith found the skin of a huge blacksnake in Codorus Township (York County), measuring 14 feet in length, in July, 1883; James Zepp and James Myers, two bark peelers (people who gathered tree bark for use by tanners) supposedly killed two black snakes, both over 14 feet in length, again in York County. Next we have the report of George Shaffer, a milkman from Mont Alto (Franklin County), who reported a run-in with a blacksnake between 12 and 15 feet in length one day in 1911.

Another article from 1917 once again refers to the monstrous blacksnakes at Big Round Top, "where 'whoppers' have been reported for years." John Rosensteel had killed an 11-foot snake, and Sol Pittenturf saw a 14-foot blacksnake. Yet another mention of these snakes – and apparently the one which quashed all further reports – came in September, 1931, when a large snake between 7 and 12 feet long was explained to be an escaped python from the home of a Maryland man. Apparently, no more stories have been forthcoming since that time.

An undated story from Blue Ridge Summit (Franklin County) tells of a William H. McAfee, who killed a 9-foot blacksnake on Rowland Hill. Somewhat arrogantly, it is said that it "is the biggest snake reported killed around that or any other neighborhood this year, or any other year."

The Lehigh Valley and eastern Berks County is another hotbed of reports of monstrous snakes, but these may not be what they at first appear. All italics in the following reports are mine; the reason for them will soon become apparent.

In 1870-71, there was a small flap of sightings from the Allentown (Lehigh County) area. The snake was always described as female (though how you'd know a snake was female without close inspection, I don't know). The serpent, a monstrous blacksnake, was said to be between "twenty-five to thirty feet in length, and the thickness of a common stove-pipe."

First, it was seen *in a meadow*, "the path made by her course looking as if a heavy log had been dragged along." It was seen by a peddler as it was engaged in crawling across a road, its tail just in view *beneath a fence* and its body stretching across the road, disappearing *through a field of rye*. It was pursued, but lost the hunters in a rocky area. In September of 1870 it was seen *in a field* with its head raised and a chicken in its mouth. Then it was seen in South Bethlehem (Northampton County) with a cat caught in its coils. The man who came across it was "palsied with fear."
A rumor surfaced from the tiny village of Windsor Castle (Berks County) that a young man by the name of Franklin Rubright was attacked by a 15-foot black serpent one evening in 1874. He fought the snake, crushing it with a club; as he dragged the limp carcass away, the snake recovered. He fought it again, but this time it wriggled out of his grasp and disappeared *into a field*.

Again, in 1887, in Lehigh County: Elias Moser of Lynn Township fought with a 16½-foot blacksnake which slithered out *from underneath his fence*. He killed this snake by throwing stones at it until it was crushed to death. The newspaper report describes his kill: "The head was long and flat. The upper part of the body was a bluish black, except two broad white bands around the neck. The belly was yellowish white." It was said that this snake was said to be a king snake, which is indeed a real type of serpent – but is the kingsnake (*Lampropeltis* sp.) we know really what was meant? Charles K. Henry and Daniel Schroeder, also of Lynn Township, killed two 8-foot snakes around the same time.

W.J. Hoffman writes of Pennsylvania Dutch superstition:

> "Occasionally we hear of black snakes found in pastures where they suckle cows, so that these animals daily resort to certain localities to secure relief

## The Mystery Animals of Pennsylvania

The so-called Valley of Death in Gettysburg - another spot frequented by giant snakes.

from a painful abundance of milk .

Some of these house and farm snakes wear crowns, and are then termed king snakes. Such were reported from several localities in Lehigh County, one of which was said to abide in a large pile of rocks near Macungie. It was seldom, however, that this golden-crowned serpent was seen; still, the greater number of residents thereabout were firm believers in the truth of the report."

Now, I should make it clear that I don't believe the 'king snake' to be anything more than folklore. However, this makes it clear that there was a native tradition of something called a king snake that was not what we *think* we're reading about.

My contention is that these Lehigh Valley reports were, like the *lindorm* (dragon) experiences in the Småland region of Sweden from approximately the same time (originally chronicled by Gunnar Hyltén-Cavallius, who was convinced of their reality) hallucinatory experiences possibly connected with epilepsy. In a 1982 *Fate* article, Sven Rosén quotes Dr. W.G. Walter, author of a pamphlet on epileptic hallucinations: "the content and structure of the experience derive clearly from the cultural and personal background of the persons concerned." Dr. Walter also notes that uneducated people often will resort to cloaking the experience in the guise of folklore. It is possible that the "palsy of fear" with which the Northampton County witness was afflicted is a further hint at an epileptic cause, and indeed this is a common feature of many of Hyltén-Cavallius' accounts as well.

Another interesting feature of note is that among the Norse, the *lindorm* is called the king of the snakes and among the Finnish it often possesses some attribute marking its sovereignty: a crown is one of those mentioned specifically. Finnish settlers would have been part of the Swedish colonizing population as Finland was part of the Kingdom of Sweden at the time. Lehigh County was just outside the northern bounds of the colonial holdings and it's not impossible that some frontier settlers made their way there.

Neil Arnold and Mathijs Kroon report that similar traditions of a king snake appear across the Netherlands. Most similarly, in Friesland, where it is called *kroantsjeslang*, the king snake wears a crown, just as in Lehigh County. And as it is mentioned that the king snake can summon lesser serpents to his aid, the Dutch variety of this monster is quite similar to the stories surfacing from Delaware City described later (and interestingly, most of Delaware was settled by the Dutch).

Schuylkill County – quite obviously a Dutch name – has some folklore about snakes. In 1906, six women on a front porch in Schuylkill Haven "were completely hypnotized by a monster copperhead snake, which emerged from under the porch and coiled in front of them." The snake was later clubbed to death by a man named Charles Detweiler.

Then, in 1917, in the same town no less, two men discovered a rattlesnake. It was only five feet long, and its rattles had been cut off (so how exactly it was determined to be a rattlesnake, I don't know), but it was "believed to be one of a number that was brought from Texas by a local soldier." The original article in *The Call* notes that one of the men was confined to bed following this event – it ascribes this to exhaustion, as it humorously points out that the man fled the scene in a great rush – but was his confinement more in line with the above-noted "palsy of fear," thus aligning this with the dragon mythology?

Another, more clearly folkloric (unbelievable, frankly), monstrous serpent encounter supposedly took place sometime in the latter years of the nineteenth century. William Johnson of Jenner Township (Somerset County) reported that individuals attempting to enter a country schoolhouse had to step over a monstrous snake whose body was a foot in diameter. Anyone who trod on its body were promptly thrown to the ground, though whether by movements of the snake or some phantasmal force, I don't know. The head and tail of the snake were never seen – a feature of some dragon tales. Most unbelievably, though, the snake was only visible on nights of a new moon, and even then not everyone could see it. Its throwing powers affected everyone, though.

This snake was making appearances at least a decade and a half later. Johnson reported that sometimes, drunkards would futilely assail the schoolhouse ouroboros with stakes and other farmyard tools. Clearly, this was The Wyrm Unvanquishable, Save for Sacnoth.

An equally unbelievable account comes from the Pocono Mountains of the northeast, where a man claimed to have sat down on a snake upwards of a half mile long.

Another Somerset County report was of a 12-foot blacksnake shot and killed as it emerged from a hole in the graveyard of Brother's Valley church. This report, too, has some similarity to several of the reports Hyltén-Cavallius gathered as many of those dragons had a direct relationship with the dead (one emerged from a grave-mound, another was the ghost of a miserly man). This snake was killed in November of 1906 and it was noted that there had been an escape of several exotic snakes from a circus years earlier. I suppose it's not beyond the realm of possibility that this was a real snake, although it's suspect.

Monstrous blacksnakes the size of logs were said to be found in the marshes around Peck's Pond (Pike County). Also hiding in these marshes were 15-foot legendary monsters called horn snakes, their tails tipped with a venomous sting like a scorpion's.

The horn snake's cousin, the hoop snake, supposedly dwelt on Buckwampum Hill in northern Bucks County. William J. Buck says:

> "This was said to be a large gray snake, haunting near its summit, that would take the extremity of its tail in its mouth and roll in the manner of a hoop with a speed equal to the fleetest horse. Occasionally, to show its power, it would direct its course towards a tree, when it would suddenly let go its tail, which was armed with a most formidable horn or sting, and let it strike into its devoted trunk to the very socket, and then continue in its progress. Strange to say wherever dead trees could be found standing over the hill, these identical holes would be hunted up and pointed out a few feet from the surface of the ground..."

Lynn Slocum killed a 9-foot blacksnake on the road between Olanta and Bloomington (Clearfield County) after he struck it with his car. A second snake also present, and an even larger snake than the first, disappeared into the tall grass at roadside. This account is given in the *Altoona Mirror* for July 13, 1929.

The Blue Rocks, near Lenhartsville (Berks County), were held to be the Devil's work – it seems Old Nick had a hankering for some potatoes, gathered some rocks up by mistake, got mightily pissed off when he found he'd done this, and ripped open his bag, scattering the boulders all across the landscape (this same event was also thought to be the origin of the Devil's Potato Patch in Lehigh County). Below the Blue Rocks was a horribly twisted and diseased tree, created in the 1780s not by the Devil, but by what may be Pennsylvania's only *bona fide* dragon.

A group of men were standing beneath a tree talking. Suddenly, with a horrid shriek, a dragon came flying over the Blue Mountains, staining the sky red as it winged its way towards the storied peak of Hawk Mountain. (This dragon, by the way, was held by some to be a gestalt entity created by the spirits of two Indian lovers, torn apart by two warring tribes, who, like Romeo and Juliet, both committed suicide. The dragon flew from their cave – called *drachelechar* or dragon's lair – to the Pinnacle of Hawk Mountain on the anniversary of their deaths. This would be the first time I ever heard of *one* ghost of *two* people.)

But after reaching the mountain, the dragon turned and flew towards the men, who, as I'm sure you'd do in the same situation, scattered and took refuge among the Blue Rocks. The dragon, thus cheated of a good meal, settled in the branches of the tree and, smoke emanating from its nostrils, flapped its wings a few times. The tree was twisted and diseased by the presence of such an infernal entity.

There's also a cave on a hillside near Virginville (also in Berks) called Dragon Cave. But of course, the Virginville cave flatly contradicts the legend of the doomed Romeo and Juliet of the Blue Mountains. But so such things go.

The will o' wisp – also called a ghost or spook light – was called a *drach* or dragon by the Pennsylvania Dutch. The mention of this tradition makes the story of the Blue Rocks and its tree more likely than a dragon myth would otherwise seem. Much is made in the story of the dragon's fiery nature, which would be expected from an encounter with ball lightning. The tree could have been blasted by an impact from the fireball.

The Poquessing Creek, marking the boundary between the city of Philadelphia and Bucks County (and around which William Penn was originally going to place the city of Philadelphia) was called by Campanius (see Chapter 3) *Drake Kylen*. Another name for the creek appearing on Swedish maps was *Riviere des Kakimons*; later writers have identified "kakimons" as a name for the pike. Possibly the "dragon" of Campanius' name for the creek was a further reference to the pike, or was it possibly a reference to something else? Perhaps the *manitto* appearing in Campanius' writings?

However, on another map of New Sweden, while the Poquessing is indeed labeled *Draak*, the name *Kikimens* (which I assume is the same as *Kakimons*) is given to the Neshaminy Creek in Bucks County. What's going on here?

Dragon Creek and Swamp, is located a short bit to the south in the state of Delaware. The creek rises in the swamp, which lies southwest of Delaware City, and flows into the Delaware River. An encounter with a gigantic serpent and another that may not be all it seemed was reported from along this waterway. In November, 1882 James Cheeseman was driving his carriage over the St. George's Causeway when the wheel of the carriage was smashed by a 20-foot long black serpent which approached along the road. Mr. and Mrs. Charles Brown were startled by the same serpent a week before.

Another man, who chose to remain nameless, was in the swamp when he encountered "what he believed to be a fallen tree" which was actually an immense serpent, each individual scale "the size of [a] soup plate." He tracked the serpent into the forest, where he found ten feet of pissed-off giant blacksnake protruding from the top of a hollow tree. The nameless fellow unloaded both barrels of a shotgun into the snake, and the slugs merely bounced off its scales but served to get the giant snake even more angry. It slithered out of the tree, and almost a hundred small heads, "evidently the young of the monster," peered out of the vacated log.

An incredible tale, and yet another that seems more similar to a dragon-slaying myth of yore than any kind of true occurrence – at least the experience of the conveniently nameless man does. The others sound more like "true" encounters, although the coincidence of a giant snake encountered along Dragon Creek along a causeway with the name of a famous dragon-slayer is interesting, to say the least.

Like the Lehigh County encounters described earlier, the nameless man's is likely the result of a *lindorm*-style experience. The 'living log' is a common feature of many of the original accounts gathered by Hyltén-Cavallius, as it also is of a number of dragon and lake monster tales worldwide. It is also a feature of the peddler's account from Lehigh County. In addition, the appearance of a number of smaller serpents is mirrored in the 1856 case of J.A. Kronholm of Småland, in which a swarm of small vipers appears in progressively greater numbers *preceding* the appearance of the dragon, in this case. This could be directly related to the Swedish accounts as Delaware City was well within the bounds of the Swedish colony along the Delaware.

Yet another Dragon Run and Swamp is located in Virginia. Dragon Run flows into the Rappahannock, which in turn flows into the Chesapeake Bay, notable for its very own monster, Chessie. Chorvinsky and Opsasnick, in their *Strange Magazine* article noted in the bibliography, give many examples of serpentine and draconic creatures seen from coastal regions of Maryland along the Potomac River and the Chesapeake.

# Chapter Ten
### Water Monsters

Most places in the world have no shortage of water-going monsters, particularly Canada, Ireland and Scotland. Pennsylvanian dracontologists (as cryptozoologists making an especial specialization in water monsters are known) have a hard time of it – for despite all the creeks and rivers that flow through the state, we're sorely lacking in that department. Traditional, long-necked water monsters, anyway.

Prehistoric life is oft-relied on when it comes to theorizing about a water monster's identity, but even that seems to be sorely lacking in Pennsylvania. Besides (comparatively) unimpressive lifeforms such as trilobites and brachiopods, some phytosaurs (prehistoric reptiles evolutionarily convergent with crocodiles) and giant salamanders have been found in York County, another giant salamander with a gila-monster like beaded texture to its skin was found near Pittsburgh, and the skeletons of the carnivorous prehistoric fish *Dunkleosteus* have been found throughout the state.

In a tally of the state's water monsters, I should mention first the oft-rumored Ogua. The Ogua was a monster, apparently known by the Native Americans to inhabit the areas along the Monongahela River in western Pennsylvania. It was supposed to have been a monstrous amphibian creature which lived in the water. Its tail would lash out and snare some unsuspecting deer or stray hunter and drag them down into the depths, never to be heard from again. It's also told that some European settlers in the Three Rivers region wrote letters about monstrous salamanders that they had killed. The fossil of a species of giant amphibian, resembling a salamander, and well-adapted for terrestrial life called *Fedexia* was discovered in 2010 near the Pittsburgh International Airport. A reconstruction of this creature can be seen on the facing page.

Pat Riley and Curt New, two fishermen from Monessen (Allegheny County) had an encounter with a monster on, as it would happen, the second day of fishing season: May 2, 1912. The two men

> "...went fishing yesterday morning in a skiff and anchored off the shore of the Page mill near the lower end of town... About an hour after

the pair had started they were suddenly attacked by an immense denizen of the Monongahela which could not have weighed less than 50 pounds at a first glance and it is hardly possible they took more than one look... the two rowed with all their strength to the shore, the river monster pursuing them as best he could. The boat landed on a muddy portion of the shore and the pseudo fishermen hurried to safety under the shelter of the over hanging cliff. The fish in the meanwhile was "a beating" it towards the bank and not seeing the mud until too late to stop, poor Mr. Fish went halfway into the bank where he stuck."

The article goes on to say that the massive catch was eaten for dinner. There's no description of what it looked like, which leads me to believe that it was either just a very large specimen of some normal species, or that this was just another fishing story, albeit one in which the big one didn't get away. In any case, other than being in the Monongahela, there's nothing to suggest a connection with the legendary Ogua; the encounter occurred near the modern I-70 bridge.

Although this story leads me to one inescapable conclusion, with which most anyone who's spent time poring over old newspapers would concur: I pine for the days when newspapers would regularly feature a run-down of the catches made on opening day of fishing season, or who's moved to what town, or who's in town visiting their parents. A simpler time indeed!

Another monster which supposedly dwelt in Pennsylvania was the monster of Wolf Pond (Dauphin County), about twenty miles northeast of Harrisburg near Elizabethville. I use the past tense because the Wolf Pond monster was seen only once, in September of 1887. At this time it appeared to a fisherman, capsizing his boat with its body and then diving into the depths of the waters. Folklorist Charles Skinner, who reported the sighting, says "The creature had a black body, about six inches thick, ringed with dingy-yellow bands, and a mottled-green head, long and pointed, like a pike's." The fisherman said the creature was nearly 30 feet long. But as the description tallies reasonably well with the northern watersnake (*Nerodia sipedon*) I think we may assume that a large watersnake was all that was seen and the fisherman exaggerated more than a little bit.

Raystown Ray is the name given to the lake monster inhabiting Raystown Lake in Huntingdon County. The first sighting of Ray was in 1994, when John R. Pendel saw a shiny black hump between 10 and 12 feet long that seemed to turn over and sink again. There were several reports in the summer of 2006 of a snaky form between 8 and 12 feet long. One night in August of 2008, Penny Foor's boat was shaken by a large wake, and one of the more interesting sightings came on August 3, when Mike Sieber saw a snake-like head sticking out of the water about 3 or 4 feet at 7 Points. It left a wake behind it, and most interestingly, Sieber noted that 15 or 20 feet behind it was a second wake.

However, despite the corpus of sightings of Raystown Ray, the fact remains that the lake is an artificial one, created by the damming of a branch of the Juniata River in 1913. It would be possible, though unlikely, for any sort of lake monster to take up residence within its waters. It is intriguing, however, given the presence of the gigantic Broad Top Snake reports to the

**A highly-dubious photograph of "Raystown Ray".**

south; one of these reports, in fact, was on the shores of the lake. Could the snake be something amphibious? *Pituophis*, the favored identity for the Broad Top Snake, is a fossorial (burrowing) species. Although remaining skeptical of the story, I too must reply with a resounding "maybe, but probably not."

Clyde Ruoff saw a monster of some kind in the Allegheny River near Smethport (McKean County) in the summer of 1950. A one-eyed creature, something with four legs and a long tail, lay in about a foot of water. Ruoff tried to catch the animal to prove his story, but it got away. Since he says it had four legs and it was in relatively shallow water, perhaps Ruoff's "monster" was just a malformed salamander?

Another runt of the lake serpent litter is rumored to inhabit the Tumbling Run reservoir near the Lower Tumbling Run Dam (Schuylkill County). This 15-foot creature was mentioned in a 2002 *Pottsville Republican* article as having been often reported in Tumbling Run's resort days around the turn of the twentieth century. As of the time of this writing, however, no contemporary resources for such a creature have been found – as such, anyway. During the reser-

Lower Dam, Tumbling Run, Pottsville, Pa.

voir's resort tenure (a short period from 1890-1911), there were nearly a dozen well-documented drowning, several of which were in the space of a few weeks in 1895. Could the numerous deaths which occurred at the lake have been transformed over the years into the tale of a "monster"? There were also a number of bizarre encounters with snakes recorded from Schuylkill Haven, only a few miles to the south, which also could have possibly contributed to the legend.

A creature of some type was reported as having taken up residence in the Susquehanna River. A journalist from Sunbury, Ken Maurer, wrote the following piece, which appeared in the *Sunbury Daily Item* in August, 2009. In part, Ken writes:

> "The other day an acquaintance who shall mercifully remain nameless came up to me and told me he read of my experience in the paper, and he was amazed because he witnessed the same mysterious sighting. His sighting was a couple of miles downstream from the area where I saw it. We discussed it at length. He felt that because of the size of it, it was a mammal of sorts, similar to a seal or otter.
>
> I felt it was a fish of some kind. After much discussion, we sort of agreed that it must be a fish because the head never comes out of the water. I have witnessed seals, otters and beavers swimming, and the head always comes out of the water somewhere along the line.
>
> Now, as to how this all started. About eight years ago, a good friend of mine told me about this "thing" he saw swimming in the river. He de-

scribed a small submarine about to surface.

Of course, I thought he was nuts. Then one evening we went fishing and the "thing" showed up. At first I thought it was a deer swimming across the river, then it turned and came upstream. When it got closer, there was nothing sticking out of the water. It pushed a wake that made waves that lapped up on the shoreline. At about 50 yards, it sank out of sight. Creepy. Over the next year or two, I saw it several times and it always sank out of sight before it got close enough to be seen clearly.

The only fish I can think of that could create this disturbance is a huge carp. I've never seen a carp act like that, but what else could it be? It's not a mammal because nothing ever comes out of the water. Between those of us who have seen it, we think it must be at least five or six feet long, which is far larger than any carp I've ever seen."

Five or six feet long? Not exactly monstrous.

Theories flew fast and furious as to the creature's identity, even though Maurer's report really doesn't detail it too well. Some believed that it was a large catfish. Flathead catfish, an extremely predatory species, are an invasive species now found in the river. Greg Misenko caught the Susquehanna's first flathead near Safe Harbor Dam (Lancaster County) in 2002. The flathead catfish was normally native to the Allegheny watershed in the west of the state, although Mike Kaufman of the Pennsylvania Fish & Boat Commission said that the species' range is apparently spreading, as it is sometimes captured in the Delaware and Schuylkill Rivers. The Fish & Boat Commission tells anglers not to throw back any flatheads they might catch.

Others endorsed the idea of its being a manatee. Manatees often lose their way and find themselves in the Chesapeake Bay; in fact, it has become nearly a yearly occurrence in the summer months. Only a month before Maurer's sighting made its way into the papers, in fact, Officer Marcus Rodriguez of the Havre de Grace, Maryland, police department saw an aquatic mammal – a manatee – nose around some aquatic weeds before swimming away, at a local marina. What makes this sighting interesting is that Havre de Grace is at the mouth of the Susquehanna where it pours into the bay.

Loggerhead sea turtles also occasionally range into the Bay and could make their way into the Susquehanna. Chorvinsky and Opsasnick (1989) mention the sighting of a loggerhead as early as 1939, and on August 14, 2009 the *Baltimore Sun* reported on a sighting of a loggerhead just off Kent Island (coincidentally, a famous sighting and filming of the Chesapeake Bay's most famous resident, the sea monster Chessie, took place off this very island).

However, few of these species except the catfish could make their way past the Holtwood and Safe Harbor Dams, so these are likely moot points.

I would like to forward another theory – a mudpuppy. Mudpuppies (*Necturus maculosus*) are large salamanders, often a foot or more in length. They are related to olms and sirens, and have external gills. They are superficially similar to the hellbender (*Cryptobranchus alleganiensis*), although they aren't closely related.

Although most of the Susquehanna's path is outside of the range of the species, my grandfather recalls that when a mudpuppy was caught, it would be hung up in a tree – a pretty cruel way to dispose of a salamander, no matter how repulsive it might be, but that's kids for you. I also remember my dad telling me about catching them in the river as recently as the 1950s and 1960s so it's possible that some are left.

Speaking of far-ranging species in the Chesapeake, Pennsylvania too has seen a number of non-natives nonetheless be found.

Powell Sherrick caught a large eel near Washington Boro (Lancaster County), in July, 1861 – "for the Susquehanna it is considerable of a monster. The skin stuffed measured four feet five inches in length, and nine inches around the head." Eels are now quite rare in the Susquehanna River, their numbers having been considerably diminished by the construction of the Safe Harbor, Holtwood and Conowingo dams.

The account of Sherrick's catch is also interesting in that it gives the price of a beer in 1861 – a mere 4 cents!

On August 19, 1868, a creature "pronounced an eel by some, a monster water lizard by others, but the prevailing opinion is that it is an infantine alligator" was being held at a drug store in Columbia (Lancaster County).

> "It was captured in the Susquehanna river near Washington borough... It is about fifteen inches long, and is a perfect alligator in every particular, except that it has no scales. We are informed by those familiar with the headwaters of the Susquehanna, that alligators of small size are quite common there, and that they derived their origin from a large one of that species placed there by the Indians some thirty years ago. The one in question must have found his way down the river to this point."

The bit about the "headwaters of the Susquehanna" is a bit odd. The actual headwaters are near the Finger Lakes in New York State, and I doubt very highly that a tiny reptile would find its way the whole way down the river to Lancaster County. It seems a bit more likely that they are referring to the region around Havre de Grace, Maryland, where the Susquehanna flows into the Chesapeake. If so, it is doubly interesting given the presence of reptilian creatures noted by Chorvinsky and Opsasnick from coastal regions of Maryland. Although, in either case, I find it extremely hard to believe that anyone would willingly introduce *alligators* into the ecosystem of an area! What this fellow was is a bit hard to determine. Obviously, since it is described as having no scales, a salamander identity is suggested. If it were a mudpuppy, I'd think the report would mention its pronounced external gills or 'horns.' Perhaps a hellbender? These certainly reach the size attributed to this 'alligator,' appearing on the whole quite similar to the giant Asian salamanders.

Another alligator was reported in the *Marietta Register* on August 3, 1889, quoting newspapers from Scranton. However, the events described seem to have happened in Susquehanna County: Hallstead and Great Bend, both named in the text, are in that county.

> "For several years boatman [sic] and others living along the Susquehanna river, between Susquehanna and Red Rock, have been interested, not to say disturbed, by a creature in the water at the latter point. The strange marine animal always made its appearance at night, says the Carbondale *Leader*, and an unearthly, weird noise, accompanied by splashing, have often wakened people from their slumber. Last fall a party returning from a harvest dance at John Dalton's was upset while crossing the stream, and two persons narrowly escaped being drowned.
>
> The majority of the party were of the opinion that the boat had collided with a log, but one or two keen-eyed ones solemnly affirmed that they had been pinched while in the water, by some submarine monster.
>
> A few evenings ago, while a dance was in progress at Michael Hagan's at Smoky Hollow, the festivities were interrupted by Bob Brown, who put his head inside the ball room and yelled: "For God's sake, boys, come down to the river quick! I've seen the whale, and he is taking a rest on the bank opposite the tannery." Every man armed himself, and rushing pell-mell to the river, the great animal was seen devouring the carcass of a sheep. As soon as he saw the men, he dropped his supper and slowly crawled into some bushes skirting the shore. The men, thoroughly excited, hastily constructed a raft from some old flood trash and pushed out into the stream. They had not gone far, when the great jaws of the monster were set upon a board in the raft, but it was its death move, for in an instant every man on the raft attacked it.
>
> The monster made a terrible shriek, splashed the water violently, and finally with one loud expiring groan, sank beneath the surface, its blood coloring the waves. Securing a rope the body was hauled ashore, and by a lantern's dim rays it was found to be that of a monster alligator. To

say that the captors were excited and delighted would be drawing it mild. In an hour every inhabitant for miles around was present. Steelyards were secured and the saurian was found to weigh 533 pounds. A physician from Hallstead, who is also a taxidermist, was sent for and he soon arrived, along with a score of townspeople. He advised sending for Dr. Crozier, a noted taxidermist, of Cornell University, who arrived and was astonished, not only to find such a specimen of the saurian kind butt [sic] that any had been found in the Susquehanna.

He gave it as his opinion that some pet alligator and that it had at some time escaped its confines, and since grown up along the marshes of the river. An old inhabitant says that he believes the late Isaac Griggs, formerly proprietor of the old National hotel, at Great Bend, once owned a pet alligator which was sent by a relative in Florida."

Alligators can certainly grow to the size described, the record-holder being over 1000 pounds. An alligator of the weight described would likely have been somewhere in the vicinity of 7 feet long.

On September 27, 1927, a city employee in Pittsburgh, George Moul, was confronted by a strange creature in a sewer under Royal Street. Moul fought with and retrieved the creature, which proved to be a 3-foot long alligator.

This account should be viewed with a skeptical eye, however, as the original article amounts to, basically, an advertisement for an exhibit of the captured alligator at Moul's home.

The following appeared in the *Titusville Herald* for August 19, 1942:

> "Three State Highway Department workers told today how they used shovels to kill a surly, five-foot alligator which attacked them. Joseph Russell of Grampian, startled to see the reptile far from its southern habitat, said the alligator was spotted in a ditch along Route 322 at Barratt yesterday.
>
> Residents of the neighborhood expressed belief the reptile escaped from carnival trucks which passed by Sunday."

In September, 1953, Reverend Edward Z. Utts was fishing with his grandchildren near McVeytown (Mifflin County) when they caught a two-foot alligator on their hooks. Reverend Utts was keeping the animal in his backyard swimming pool.

The Pittsburgh area seems to be a hotbed of alligator sightings, however, the status of Moul's report notwithstanding.

In October, 1998 a three-foot alligator was found swimming in the Kiskiminetas River just north of Schenley (Armstrong County), and another was caught in an alley in New Kensington (Westmoreland County) the following year.

Two of the wayward reptiles were encountered in the Beaver River over the course of a few days in the summer of 2002. One was on August 2, when three fishermen had their bait eaten and lines snapped by the jaws of a "prehistoric-looking" animal: when captured it proved to be a two-year old alligator nearly three feet long.

The next was encountered only two days later. Several individuals saw an alligator in the river near the Townsend Dam, in New Brighton; but like so many fishermen, these witnesses were left only with the tale of "the one that got away."

The story of the Beaver River alligators doesn't end there, however: three years after these two encounters, in August of 2005, there were two more. And ironically, one of these again took place near the Townsend Dam when Tim Rodriguez found a three-foot alligator.

Only a few hours later, another alligator was found. (The month before, another Tim – Tim Coleman, of Chicora, in this case – discovered a four-foot alligator lying alongside a stretch of Route 422 in Butler County.)

An alligator found in the Allegheny River near Tarentum (Allegheny County) a month after the 2005 Beaver River encounters proved to be an escaped pet. But the status of a two-footer found in September, 2006 in Frick Park, just outside Pittsburgh, was never determined for certain.

Speaking of Three Rivers area alligators, one was suspected of starting a fire in Lawrence County. In March, 2009, a rundown former school burned to the ground (the resident, Brian Simpson, reportedly had a "menagerie" of more than 70 animals in the building, many of which died) and a six-foot alligator was rescued from the blaze. Simpson said that the fire was sparked by a heater knocked over by the alligator in question.

Some of the sightings from the Pittsburgh area may have been of hellbenders, discussed earlier, a rapacious breed of salamander reaching lengths of up to 2½ feet. Hellbenders, though rare, are definitely found in the area, and in fact the species once, ironically, received the name "Allegheny alligator". Undoubtedly, though, many of the reports are of individuals kept as pets and later released.

On the other end of the state, the Philadelphia area is another hotbed of reports. Three alligators were captured in Bucks County, north of the city, from 2007-2008. Two 'gators measuring three and five feet long were caught in the Pennypack Creek in 2007, and one was captured at a construction site in Northampton in July, 2008. Yet another was roaming the area of Horsham (Montgomery County), and that reptile was never recovered. Three months later, in October, boaters on the Schuylkill River saw an alligator. Later in the week, the four-foot reptile was captured in the Strawberry Mansion district of Philadelphia.

In early September, 2009 an alligator was captured in the Log Basin, a pond in Stacy Park in Trenton, New Jersey. The Log Basin is along the Delaware River, just over the New Jersey state line and just north of the Trenton Rapids discussed later. The article appearing in *The Trentonian* newspaper called it the "Loch Stacy Monster," needlessly likening a simple alligator to Scotland's elusive beast.

"There might a second, larger alligator in the Trenton "loch" where a 4-foot gator was captured yesterday, wildlife officers and neighborhood residents said. "That's not the alligator we've seen over the past several weeks," resident Jesse McCray said after yesterday's capture. "We were down here on Tuesday and saw an alligator with a much larger head, at least three or four times the size of that one." City Animal Control Officer Jose Munoz, an 11-year veteran, shares the two-gator theory: "I saw it myself before we got it in this cage and it looked much bigger. Hopefully there is not another one in there," Munoz said.

But, he added, "quite a few people are saying that this is not the alligator that they have seen. Once we transfer this alligator into another cage, we're going to take this one back out and reset it —just to be on the safe side." In light of all the talk of a second gator, technician Kim Tinnes of New Jersey's fish and wildlife division also said the state isn't taking any chances.

"We're going to reset the cages just in case," Tinnes said. "We've heard some conflicting reports from people who have seen an alligator. It would be our worst nightmare if there were another alligator in there."

Trenton's Log Basin, the pond in riverside Stacy Park that has been home to at least one gator for at least three weeks, attracted dozens of curious kids and adults after word of yesterday's capture got out. And all the speculation about another mysterious monster remaining in the pond in Trenton's Island section put some in mind of Scotland's Loch Ness, a tourist trap said to host a prehistoric beast called "Nessy."

Darlene Yuhas, spokeswoman for Jersey Department of Environmental Protection, called the second gator theory a crock, saying from her office in downtown Trenton that there's no real indication of another reptile out there. But Island resident Bruce Berenson, who has filled his camera with Loch Stacy Monster sightings in recent days, thinks an even bigger alligator remains in the urban pond.

"I've been down here about three or four times. My first guesstimate was that this one looks small compared to what I saw before," said Berenson, pointing to the little gator angrily hissing in its cage. "There's no way to be sure but my first instinct was that the alligator seemed pretty small. Mind you, I saw this (alligator) through binoculars while it was sprawled across a downed tree. Still, I think I had a pretty accurate view of its size."

Berenson suggested that Trenton may be headed for an alligator sequel. "You've got to wonder if there were two in there." Then pond visitor Aunia Gurley suggested there might be a whole family of gators out there. Yesterday's captured gator is "probably one of the babies," she said."

Just a week later, on September 10, an article appeared in the *Philadelphia Inquirer* detailing the capture of a good-sized alligator in Allentown (Lehigh County). The day previous, a pedestrian saw an alligator lying in a busy park on the banks of Jordan Creek. Police and animal control officers wrestled the reptile out of the creek, and found it was a 10-15 year old individual, 6 feet long and 130 pounds. The Allentown alligator was later sent to a zoo near East Stroudsburg (Monroe County). As luck (or the coincidence that so plagues researchers into the unexplained) would have it, the zoo it was sent to was one Mindi and I had visited on our honeymoon, although that alligator wasn't there yet.

A report appearing in the *New York Times* in 1922 noted that a 12-foot shark "said to have been of the man-eating variety" was found and killed in the Delaware River by Joseph Fletcher of Tacony (in north Philadelphia). Although a number of shark species are found in the Delaware Bay (including dangerous varieties like the hammerhead and thresher), only one would range into the river itself with any frequency. That is the infamous bull shark, which often ranges into fresh water. The bull shark is a dangerous species, and is probably responsible for many of the documented shark attacks. The bull shark is one of the favorite suspects in the infamous 1916 shark attacks in Matawan Creek, New Jersey (the saga which inspired Peter

Benchley to pen *Jaws*). "A shark of seven feet length" was captured off Philadelphia in June of 1735 and another, nine feet long, was captured near Windmill Cove (Windmill Island is in the Delaware River at the end of South Street in central Philadelphia).

Just after Christmas in 2006 (December 29), employees of the streets department of Punxsutawney (Jefferson County) were working near the corner of South Gilpin Street and Cypress Street. Their attention was drawn to a large object near the banks of the Mahoning Creek. Walking over towards the shore, the employees saw that it was a large fish. Drawing even closer, they noticed that it was a three-foot bonnethead shark (also called a shovelhead shark), a species similar to the hammerhead with shorter, thicker and curved protuberances on its head. One can imagine the surprise of the employees, as Punxsutawney is far inland, and quite a distance from any sizable body of water.

We return to the Delaware River once more for a discussion of several encounters with whales of various species recorded from the vicinity of Philadelphia. "Thirty od years ago [sic] A whale was Exhibited in York in the Barn of Samuel Spangler, got in the Delaware river. 30 feet long," reads the horribly error-ridden caption to a drawing by Lewis Miller. The whale's body was displayed on a wagon. While the date is unknown, it is from prior to 1850.

Rhoads (1903) says that in 1862, Edward D. Cope described the black whale or Atlantic right whale, *Balaena cisarctica*, based on a specimen captured in the Delaware. According to *Watson's Annals*, Phineas Pemberton, who built the Bolton Mansion (a famous Pennsylvania haunt), wrote that a whale of indeterminate species was seen in the Delaware River near Trenton Falls as long ago as 1688 – only seven years after the city of Philadelphia was founded. Trenton Falls, despite the name, are actually rapids which are located underneath Trenton's Calhoun Street Bridge which crosses near Morrisville (Bucks County). A chase was mounted for a whale and her calf seen in the Delaware in April, 1733, but they managed to elude their pursuers. Another "huge potentate of the scaly train", as contemporary accounts described a whale, was taken off Chester in 1809.

In December 1994, the *Philadelphia Inquirer* reported on the 30-foot "Waldo the Wrong-Way Right Whale" (*Eubalaena glacialis*) which was seen in the Delaware. Wildlife officials managed to divert Waldo towards the south, but the next year he was back and was beached at an oil refinery in Pennsauken, New Jersey before being returned to the water. Waldo's wanderings were straightened out, and he was last seen in the waters off Canada.

Another whale, this one a beluga (*Delphinapterus leucas*) called Helis, was reported as far north as Trenton, New Jersey. Helis made his way into the Delaware Bay for about a week, but by April 29 he was back in the river and was seen near Burlington Island. Here he turned again, and the next sightings of the white whale took place at the Walt Whitman Bridge to the south. There were also unconfirmed reports of the beluga from the Schuylkill River. By May, Helis was out of the river and back to his normal territory in the St. Lawrence River of Quebec.

Seals, too, are recorded as having ranged up the Delaware. The earliest *recorded* sighting was in 1824 off the Repaupo Floodgates in New Jersey, across from Chester Island; thereafter, seals were reported at the Trenton Falls in 1861, 1864, 1866, 1870, and 1877, each one in the winter; eight more were reported at Trenton in the winter of 1878-1879 alone.

One was captured in February of 1870 near Bristol (Bucks County), and on October 20, 1901, the *Philadelphia North American* reported that William Hill and Joe Springard had captured a seal near Chester (Delaware County). Dr. C.C. Abbott wrote in 1880 that he felt that "in severe winters [seals] are really much more abundant in the Delaware River than is supposed"; and the roll of sightings seems to support his view.

While we're on the topic of odd sightings in the Delaware River, something was reported in October of 1921 which appeared, of all things, to be a terrestrial cephalopod of some type. Since the mass hysteria created by the wave of sightings of the Jersey Devil in 1909 was still fresh in the minds of residents of the region, the creature inevitably received that moniker. The encounter actually occurred in New Jersey, near Gloucester City:

> "The creature, which has eight flexible arms like the tentacles of a devil fish, is able to move about by means of the supports. When seen by the Italians it was moving in ungainly style toward the beach, near Gloucester. As the fishermen ran toward it the creature turned and opened a well developed beaklike mouth. Its small eyes glowed wickedly.
>
> The fishermen threw a heavy net over it and captured it."

Despite its apparent nature, the animal was declared to be "undoubtedly" a reptile by a man named Frank Kelly. It was supposedly sent to a laboratory in Trenton to be examined.

Octopi are, indeed, capable of movement on land, although this bizarre animal – assuming it was a real report and not just another of the overly-sensationalized semi-fictional accounts of the weird which proliferated in older newspapers – seems to have had a more "upright" gait than a traditional octopus would. It seems, for all the world, to be like some monster out of a Japanese monster movie.

If urban legend is believed, some monstrous fish inhabit Luzerne County. Some of these supposedly dwell in the depths of Harveys Lake in the north of the county, and an old-timer living in Exeter Township, north of Pittston, said that he caught "something that was half lizard, half fish. It had razor teeth that bit through the line." Another monstrous fish – this time the ubiquitous monster catfish – supposedly lives in Blue Heron Lake (Pike County).

These fanged fish, anyway, could just very well be an urban legend that's sprung up since the advent of the northern snakehead fish (*Channa argus*), an introduced species which had made their way to a park pond in Philadelphia by 2004.

They've since moved into the Schuylkill and Delaware Rivers, and they are predicted to make their way into the Lehigh River. Snakeheads are an amphibious species and their extremely voracious nature has led to a campaign against their advances in Pennsylvania.

They also could have been based on the catfish, many species of which can move about on land for short periods of time. It is not inconceivable that a catfish's long 'whiskers' could be mistaken for long teeth.

Of course, they could just be tall tales, too.

Another fish of a less monstrous kind, but still a stranger to Pennsylvania's shores, was caught in the Conestoga River near Brownstown (Lancaster County) in 2009:

> "Eric Laubach and Steve Bergstrom both thought there was something fishy about the "sunfish" Eric's 5-year-old son, Jake, hauled out of the Conestoga River Sunday evening.
>
> "We were catching some nice sunnies, but this one was a lot bigger than the others," Bergstrom said.
>
> Holding the 12-inch-long fish in his hand as he prepared to remove the hook from its mouth, Laubach got a look at the fish's jaws.
> That's when he saw teeth.
> "I knew Steve used to have piranhas, so I said to him, 'You gotta look at this, I think it's a piranha,' " Laubach said.
>
> Bergstrom seconded his friend's identification, but noticed "that thing was a lot bigger than any piranha I ever had."
> Based on the shape of the fish's teeth and its overall size, however, it most likely is a red-bellied pacu.
>
> Piranhas have a protruding lower jaw lined with triangular teeth used for tearing apart flesh.
> The lower jaw of a pacu is flush with its upper jaw and its teeth are closer to square than triangular. Pacus will eat anything, but prefer fruits and vegetables, according to the state Fish and Boat Commission's Web site.

And while a 12-inch-long piranha would be about as big as that species is known to grow, a foot-long pacu is a common length for that fish, which can grow to more than 25 inches. Pacu or piranha — neither one belongs in the Conestoga River. Both fish are native to South America, but are commonly kept in aquariums in North America.

Red-bellied pacus have been caught in the Conestoga River before. Officials with the Fish and Boat Commission have said in the past the tropical fish likely were pets released into the wild by their owners — an illegal act in Pennsylvania. Thinking the fish was a piranha, Laubach and Bergstrom didn't feel comfortable returning Jake's catch to the Conestoga River behind Bergstrom's house on Riveredge Drive in Manheim Township.

So they took it into the house and put it in an aquarium kept by 5-year-old Wells Bergstrom, who was fishing with Jake at the time the fish was caught. When the men did some research on the fish, they found a notice on the Fish and Boat Commission's Web site that stated such fish should be removed from the wild when caught.

"We called the Game Commission (Monday) and they told us not to put it back, too," said Bergstrom's wife, Tonya. The fish died in Jake's aquarium Monday, and Bergstrom was considering having it mounted by a taxidermist.

"It sure is going to be something to talk about for a long time," he said.

I find it interesting that the article says this was not the first capture of the species in the Conestoga.

Another pacu was caught on September 17, 2009 by George Horvath, a carp fisherman, in the Delaware River near Trenton:

"I caught a 14" 2 lb 4 oz male red bellied pacu (piranha) in the Delaware River near Trenton," Horvath said in an e-mail. "That's the second one that I caught there. [On] 7/16/2004, I caught a 12" 1.5 lb female pacu (parapatinga) in the same spot on the same bait, my homemade cornmeal carp bait."

Finally, in August, 1975, Peter Evangelidis and his girlfriend (now wife), were walking along the Delaware River and approaching the Ben Franklin Bridge when they saw "a bunch of black inner tubes or tires floating down the river, about 30 yards out" moving against the current of the river. Suddenly "this sleek head of an animal that should not have existed sprung its head out of the dark river no more than 30 feet from us." Evangelidis wondered whether the Chesapeake monster Chessie could have made it through the canal cutting across Maryland and Delaware to Philadelphia. But, as we've seen, there's any number of things it could have been. 1975 had seen reports of whales and sharks, anyway.

# Chapter Eleven
*The Tinicum Swamp Cat*

As previously mentioned, the bobcat, the only wild feline native to the state (barring a resurgence of population by the Canadian lynx), is reported to occur in a long-tailed morph at times. This chapter will attempt to tell the story of these long-tailed wildcats, beginning with one of the most famous of these. The article on the find in the *Bucks County Times* (January 20, 1922) is exceptionally thorough.

For years, the inhabitants of the townships around the state game lands in the Tinicum Swamps of that county had been terrified by nightly howls, yowls, and shrieks. The article mentions that some residents of the towns were afraid to leave their homes at night. Daniel Trouts' son had killed two similar long-tailed wildcats in the vicinity sometime around 1919 or 1920, and one had also been killed about 1840 on Spruce Hill, a knob of land about a mile south of Chalfont. In the modern day, this would be just outside the Philadelphia metropolitan area. The article also notes that the Chalfont wildcat "might have been a bobcat."

A 16-year old boy named Tunis Brady set out to capture the nocturnal yowler, and managed to track two animals to a rocky outcropping near the swamp. Frustratingly, no less than three times the animals stole his bait, sprung his traps, and avoided capture. On January 16, 1922, Brady managed to capture the male of the mated pair. The female hid herself elsewhere in the rocks.

As the description of the specimen in the news article is much more thorough than any I could write, I shall quote it in full:

> "The animal... has strong and powerful claws and teeth, and its head is large in proportion to its body. It apparently has not an ounce of surplus flesh, being sinewy and wiry, and yet it weighs eight and a half pounds. Its length from the tip of his nose to the tip of its tail is 30 inches. Its body length is 20 inches and it stands 13 inches high. Its front legs are 7 inches long and its back legs 13 inches. Its head measures 11 inches in circumference the broad way and 13 ½ inches the long way. It measures 3 inches between the ears and 7 inches across the

ears. Its body is 12 inches around just behind its forelegs and 14 ½ inches around the centre. Its tail is scant 11 inches long, thick and inclined to be bushy, which distinguishes it from the domestic cat, which has a long tapering tail."

Similar cats were described in a later article by Elizabeth C. Cox of Holicong (Bucks County) as being native to the Chestnut Ridge area in Fayette County as well as in the Blue Ridge Mountains, a spur of the Appalachians leading through southern Pennsylvania near Gettysburg (Adams County) and through Maryland and Virginia. Two were supposedly killed sometime previous to 1840 near Roulet (Potter County) by Burrel Lyman.

John P. Swope, a famed trapper and hunter from Huntingdon County, wrote in 1909 that some hunters had treed a wildcat "striped like a tiger and [with] rings around its tail like a coon." While physically it seems to be a match for our long-tailed bobcats, Swope notes that it was "9 feet long from the nose to the tip of its tail" – a feature which drags these fairly small wildcats into the realm of late-surviving cougars.

In February, 1924, James D. Geary, a state game warden, shot and killed an animal called by the article a "Maine tiger cat, said to be a cross between an Angora cat and a raccoon" – presumably in appearance, since a raccoon and a cat cannot reproduce! This, along with its weight of 11 pounds, makes it at least possible that this was a better-fed specimen of the Nockamixon wildcat's species. This cat was killed in the Blue Mountains near Allentown (Lehigh County), only about 16 miles from where the Tinicum Swamp specimen was killed.

In 1934, a story appeared in the *Pennsylvania Game News* entitled "A Wild Tame Cat?" describing the killing of a cat that was

"...37 inches long from the tip of his nose to the end of his tail and weighing 13 pounds. The head was large, the face was flat and blunt, with long whiskers, protruding tusks, and sharply pointed ears, which had long black tufts on the tips. The shoulders were broad, the body stout, the tail long. The color markings were light yellowish brown on his stomach, with dark spots. The back was dark brownish town, with dark stripes. The tail was about eight inches long, dark brown with black rings."

It's not too clear whether the story was intended as fact or fiction, but the scene of the yarn was supposedly in the mountains near the old logging town of Driftwood (Cameron County). Aside from the "protruding tusks", this seems to be describing another specimen of the animal killed in the Tinicum Swamps.

The latest sighting I'm aware of (though whether because these creatures died out or because they were called other names in more modern times) was in September of 1951, when Lynn Wykoff trapped one of these animals south of Wharton (Potter County). Wykoff's father named the animal, which had a foot-long tail, Bertha. Why? "I once knew a woman by that name who had a temper just like this animal!"

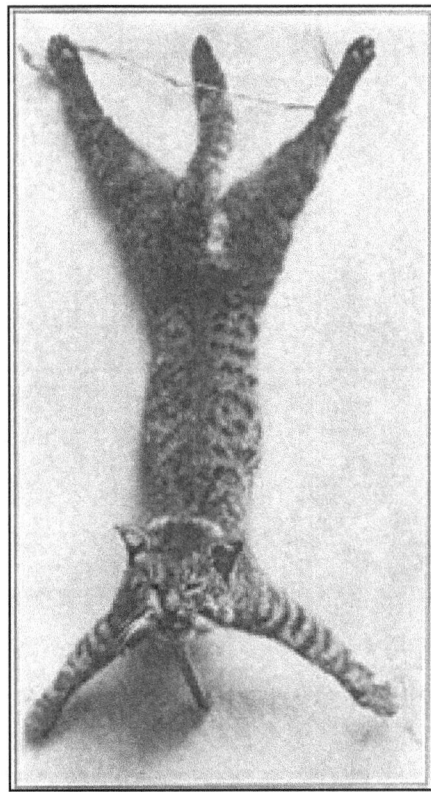

The Tinicum Swamp Cat, as pictured in the *Bucks County Times* (January 20, 1922).

It is, I suppose, possible that some of these long-tailed bobcats could have been lynxes, which have a longer tail than bobcats. The last known Pennsylvania lynx was killed around 1923 in Tioga County, although the species still occurs, albeit rarely, in the Adirondack Mountains of New York. *The History of Bucks County* says that "two or three" bobcats had been killed in the 1860s in Rockhill and Haycock Townships, but does not give examples (Rockhill is presently divided into two townships, and I don't know which one is meant). Both are in the vicinity of where the Tinicum Swamp Cat was killed. Apparent sightings of wildcats of the Tinicum variety exist in New York and New England, but also apparently come from southern Illinois (where they are known as 'wood cats' or 'timber cats'), which is outside the range of either the bobcat or lynx (although it's just outside the bobcat's range, so it's not outside the realm of possibility some could exist there).

# Chapter Twelve
## The Yardley Yeti

The first sightings of something odd in Bucks County (well, I should clarify. Not the first odd thing in Bucks County – not by far! – the first sightings of something *dog-like* and odd) started coming in, in the fall of 2005, surfacing from suburban Lower Makefield Township. Rather unfortunately for the poor animal, it was dubbed the Yardley Yeti by the media, although 'yeti' is a name with a completely different connotation among cryptozoologists. This Yeti, however, was described by Chief Ken Coluzzi of the Lower Makefield Police Department: "Some called it a cross between a dog and a hyena. Others said a wolf dog, and others said it was a sick looking foxlike creature. Others said a coyote."

Just before Halloween that year, appropriately, author Jonathan Maberry and his wife were at an art gallery in New Hope when a mysterious creature trotted through the parking lot. They managed to snap a few pictures of the animal before it disappeared. Many people seemed to agree that it was a red fox with a severe case of sarcoptic mange, but the fact remains that in some of the photos, its snout looks too short and blunt for a fox. The veterinarian consulted by Maberry seemed to agree. It seemed to doubtless be a mangy *something or other*, but exactly what, the jury's still out.

It seems that for most of the cryptozoological community, the story stopped there (cryptozoology is plagued by this sort of thing – one fairly explainable event takes place, and then nobody pays attention to any further sightings of the thing). That didn't stop the Yeti, though, because in June of 2006 a man and his wife driving through Croydon encountered a creature they thought was the infamously-named beast. In the autumn of that year, Don Polec, a journalist with the ABC affiliate in Philadelphia, had a close encounter with some kind of creature while driving along Eagle Road in Upper Makefield Township:

> "The best I can do is say it was a quadruped, grey in color, about the size of a large German Shepherd, skinny, really ratty looking, body parts that seemed oddly out of proportion, too large for a fox, legs too long for a coyote or dog, head too small for its size, and REALLY in need of some grooming and big dish of Alpo."

Only a few days after Polec's encounter, on October 14, a woman and her daughter driving along Pine Road in northeastern Philadelphia encountered a creature, "brown, almost hairless, thin, pointed ears, and large back legs."

Things got weird in December of 2007, nearly a year after the original encounters. Most every researcher into the unexplained will be familiar with how weird events seem to "follow" people around. Jonathan Maberry, again, was driving down Byberry Road in northeast Philadelphia (not far at all from Pine Road, site of the previous encounter) when he came across another creature of the doggish sort, similar to the one he photographed in 2005, but larger, leaner and gray-white in color.

I would note that it is interesting that the northeastern Philadelphia accounts begin only *after* the Philadelphia State Hospital (Byberry), a mental asylum abandoned since 1990, was torn down in June, 2006. The asylum was more-or-less centrally located between Pine Road and Byberry Road; Croydon is also fairly close to where it stood. Were these southern accounts of the Yardley Yeti of something that had previously been taking refuge in the abandoned building or miles of tunnels beneath it? Two wallabies, a male and a female, had escaped from a farm in Upper Dublin (Montgomery County) in 2001 and were never recovered. Could ill-fed, mangy wallabies, descendants of these two, be responsible for some of the Yeti sightings? Only a short time before this writing, on July 18, 2010, an animal that had all the appearances of a hairless coyote made its appearance on the outskirts of a wooded area along Gibson Boulevard in Steelton (Dauphin County), just outside Pennsylvania's capital city of Harrisburg. The Steelton animal, though hairless, didn't have the typical scabrous appearance of mangy animals. In fact, it appeared to be quite healthy aside from being bald as an egg. An anonymous witness also wrote that they had seen this animal, or a similar one, four days previous along the Conodoguinet Creek (Cumberland County).

Hairless canine animals are not exclusive to Pennsylvania, though, not by far. Since the appearance of the first one that made national news, the one found near Elmendorf, Texas in 2002, a number of them have appeared throughout the central United States. Some have been recovered from Oklahoma and Tennessee, especially as well as several more from the Lone Star State – as well as from Maryland in the east. Unfortunately, these hairless canines are almost universally called "chupacabras," though the animals surely have no relation to that Latin American vampire beast save for an apparent penchant for exsanguination.

The Steelton Bald Dog

# Chapter Thirteen
*The Ape-Boy of Chester*

One of the oldest urban legends in eastern Pennsylvania is the tale of the Ape-Boy of Chester (Delaware County), also known as the Nature Boy (no, it's not Ric Flair). An intolerably ugly red-haired boy was born in the city, and so tormented was he that he fled into the swamps around the city, gaining a wild nature and, eventually, appearance. Over time, with the spread of urbanization, the swamps receded – now, only the John Heinz Wildlife Refuge remains and it is presumably within that marshy bottomland that the Ape-Boy now dwells. The Wildlife Refuge is along the banks of the Darby Creek (we'll touch on that later), bordering the town of Lester and the property of the Philadelphia International Airport. And though you may hear it referred to as the Tinicum Swamps at times, it shouldn't be confused with the locality where the Tinicum Swamp Cat, described in a previous chapter, was shot.

In 1963, Jack E. Boucher suggested that the Jersey Devil was actually a deformed boy, who was kept in the house until Mrs. Leeds (the Devil's supposed mother) could not care for it anymore. Once this came to pass, the malformed offspring fled into the swamps, raiding the occasional farmstead to get food. The poor unfortunate was long dead, but the story lived on. The similarity of this version of the tale to the tale of the Ape-Boy is intriguing – the Pennsylvanian tale also predates the American Revolution (the Devil was supposedly born in 1735). That said, it resembles in part too, the legend of the Delaware werebeast called Red Dog Fox, another legend from the same general area attached to a red-haired boy and another apparently of Revolution-era vintage.

At any rate, the legend of the Ape-Boy seems to remain just that until the early years of the twentieth century. In November of 1906, some "wild and uncanny thing, resembling a gorilla" was prowling around Morton and Springfield Township. This beast, though, had an unusual attribute or habit – it appeared only to blacks. To me, this sort of limitation of witnesses implies that the creature was more folklore than fact; or that it was an outgrowth of the rampant racism of the era, wherein whites felt themselves "too civilized" to see such a monster.

> "It walks on its hind legs and makes its appearance about dusk...
> one negro says it growled at him a few nights ago, and that the growl

> was sufficient inducement to him to run all the way home.
>
> Another asserts he shot the beast, but that the bullet bounded off as though it had struck a chunk of armor plate."

That last line calls to mind another bullet-proof creature, reports of which were rampant in a black neighborhood of Mobile, Alabama. This was the so-called Frankenstein of Fisher's Alley, reported over a two-day period in January, 1938. The creature was variously described as being anywhere from the size of a cougar to the size of a *Tyrannosaurus rex*. The monster proved to be nothing more than an extremely large otter. This creature, though, was to spawn an urban legend of its own as the next month reports of a six-armed gorilla with a panther's head, known as the Goon of Guillemard Street, surfaced from neighboring Pensacola, Florida.

Back to Pennsylvania, though – the next month the wandering ape appeared near Darby. Gone were the implications that only blacks saw the monster; it was now thought of as a baboon, not a gorilla. It was also implied that the sightings were likely the product of some trickster. Many people were arming themselves, and the article notes that Frank Carr set a number of traps in the woods where the "baboon" was said to have been seen, and that they were broken and the bait devoured.

> "Those professing to have seen the wonderful animal assert that it sometimes goes upright like a man and then dashes along on all fours with marvelous speed maintaining a queer galloping gait. They furthermore feel certain that it has a coat of dark hair but that is not considered remarkable as the weather is cool."

The last line seems incongruous. Perhaps that baboon in the suburbs of Philadelphia isn't so strange after all – at least its *furry coat* makes sense. A rumor was flying about that the Philadelphia zoo was hunting down an ape which had escaped, but that theory was disproved. Of course, as the monkey-keeper suggested then that the ape was "of native Delaware County stock," perhaps we should question his credentials. That said, the movement and gait of the animal certainly sounds baboon-like to me.

Of course, the Darby Creek flows by the town. Since the airport wasn't built yet, the swamps were much more extensive at the time, dominating Hog Island and reaching upward along the southern reaches of the Schuylkill River. The swamps also continued eastward into what is now considered South Philadelphia. At this time, the Ape-Boy would've had a lot more territory to hide himself in.

The *Chester Times* for September 6, 1919 carried a highly fictionalized account of the sighting of a creature called Wild Man of Leiperville the night before. Leiperville no longer exists as a separate entity, but is lost in the sea of suburbs between Chester and Philadelphia.

> "A frightful looking creature, with a haggard face and a hang-dog look and with the hair on his head and partly draped figure long and shaggy... They saw the wild man with their own eyes and they felt the

The mansion of William H. Furness, who had two orangutans and a chimpanzee around 1914.

Is the screaming woman of Fort Mifflin actually the Ape-Boy?

pierce of those melancholy wails *oo-oo* and *Mag-Mag*. He screeched and whooped and at times cried like a baby..."

The story as a whole was a melodramatic tale of a man gone wild and the forbidden love he held for a widow, Mrs. Maggie Millett. The wildman was encountered by Bertrand Johnson and Armstrong Henderson, who were fishing Crum Creek. But the story was based on a real event, since on September 11, the *Times* ran another story about the Wildman. As had been suspected for several days, it said, a Leiperville man named William Worrell was the monster. Others, however, claimed that Worrell was not the culprit.

What exactly is Chester's most hirsute resident, should he exist? Many cryptozoologists would suggest a Nape, or North American Ape, a putative identity put forward by Loren Coleman to explain a great number of sightings of creatures, more ape-like than the more-or-less human-like Bigfoot, reported from marshy areas and riverbeds, mainly in the southern United States. Apes are notoriously poor swimmers – a survey of the records of zoos and sanctuaries worldwide discloses a great number of reports of these animals drowning. However, a few reports exist of incidents in which apes were seen swimming. A 1900 report told of an orangutan stranded on an island in the Mississippi, and in 1926 a chimpanzee reportedly swam the Rockaway River in New Jersey. Such reports are not uniquely North American, however. Primatologist Vernon Reynolds reports on several chimpanzees seen swimming in the Benito River (now the Mbini River), in what is now the country of Equatorial Guinea. Reynolds believed that they were not actually chimpanzees, however.

A number of reports collected from Southern waterways also seem to describe distinctly baboon-like animals inhabiting much the same sort of habitat, such as the creature encountered by Reverend Harpole near Mount Vernon, Illinois or the baboons rumored to have lived along Texas' Trinity River in the 1980s. The second set of reports of the 1906 ape, as long as they were recorded faithfully, do indeed seem to be of a baboon or something similar.

Philadelphia was the first American city exposed to the gorilla, at least the lowland variety thereof, via a skeleton sent there in 1851. Live specimens of this greatest of apes, though, didn't appear stateside for forty-six years. The lowland gorilla also displays a distinct reddish-brown patch on its conical head, and some also display similar patches on their bodies. This is interesting since the original Ape-Boy was said to have red hair. Could there have been weirdly out-of-place lowland gorillas seen in Delaware County, and could the legend have been invented afterwards as a way of explaining away the sightings? Or perhaps the Napes display a similar coloration?

To be fair, the Wildman of Leiperville seems to be much more humanoid then the earlier series of ape sightings, but that can be easily explained as we know that the first story of the Wildman seemed to be a fanciful version of what actually happened (much like what Henry W. Shoemaker was known to do at times). So who's to say that the Wildman wasn't much more ape-like than it's described?

An interesting possibility centers around the haunted Fort Mifflin. Elizabeth Pratt, a woman who emits chilling screams, is reputed to be one of the ghosts haunting the Revolution-era fort, which borders (and is nearly engulfed by) the Philadelphia airport. However, some auditory witnesses to the screams feel that they may actually come from the swamps behind the Fort – the swamps which would happen to be part of the John Heinz Wildlife Refuge. Are the screams of Mrs. Pratt actually those of the Ape-Boy?

But what if these sightings weren't anything unknown, but a genuine ape? It is known, after all, that great apes were indeed present in Delaware County – Dr. William H. Furness of Wallingford was an early adherent to the theories which eventually led to the research which taught various apes to communicate through sign language. He had "two orang-outangs [sic] and a chimpanzee, in an apartment in his greenhouse, that he has taught most marvelous things, and which seem to bear out his theory that they possess an intelligence that can be taught to think and speak." Furness lived in his familial estate, which is no longer there but stood on the grounds of what is now the Furness Free Library. His apes were present around 1914.

If the Ape-Boy is, indeed, indicative of a Nape population, we would expect them to display the same swimming ability as others of that hominid group, and possibly on occasion to cross the Delaware into New Jersey. Indeed, there are several reports of apish entities from the riverside counties of that state. These counties are similar in terrain to what the Tinicum Swamps *used* to be – but urban sprawl hasn't touched southwest Jersey quite as much as Pennsylvania and it provides *much* more swampland in which to hide, so it is possible that the Ape-Boy population may have left our shores to take up a more permanent residence in New Jersey. Not that the Pennsylvania apes are entirely gone – as recently as 2004, the Pennsylvania Bigfoot Society logged a report of a bipedal hairy man from the outskirts of Brookhaven. Brookhaven, coincidentally, is just a bit upstream along the Crum Creek from the bridge where the Wildman of Leiperville was seen.

Many researchers have noted how 'Jersey Devil' seems to be a blanket term covering everything unexplained that happens in New Jersey (and sometimes beyond, as it turns out). True to that maxim, the first report of what could have been some trans-Delaware Ape-Boys came on January 20, 1909, during the midst of that infamous week of Devil sightings and the same day that Daniel Flynn saw an alligator-skinned man near Leiperville (see later). Mrs. Mary Sorbinski, a resident of Mt. Ephraim Avenue, heard a commotion in her backyard and emerged, armed with a broomstick. She found her dog in a horrible creature's "vise-like grip." She swung the stick at the monster, which uttered sounds that were said to be a cross between a hyena's snarl and the hoot of an owl, dropped the dog, and rushed past Mrs. Sorbinski. She screamed and alerted the police, who later tracked high-pitched screams to Kaighn Hill, where they opened fire on the beast, which vanished.

Given the speculations about Napes' swimming abilities, it is intriguing that both of these sightings came in areas with easy access to the Delaware River. One wonders if when the officers fired at the 'Devil' it didn't so much vanish as simply jump into the water.

At 3:00 AM on the morning of February 22, 1924 a milkman named William Bishop was making a delivery around Sixth Street and River Drive, in Westville (near the mouth of the Big Timber Creek) when his horse neighed an alarm. He saw a

> "...tall, narrow [thin?] man about seven feet in height with hair hanging down longer than the tail of a horse. The hair was a sort of gray, he said, and the object had a burlap bag tied around its waist. The "devil," or whatever it was, took two bottles of milk and a package of cottage cheese from the milk wagon..."

Its booty in hand, the creature disappeared towards the creek. Charles Dempsey, also of Westville, found tracks in his yard which Bishop said matched tracks the creature had left.

A driver near Salem in 1927 said that he had just changed a flat tire on his car when his vehicle began to shake. A creature "that stood upright like a man but without clothing and covered with hair" had its hands on his car.

A black, hairy creature that exuded a foul stench was reported from the forests along the banks of the Rancocas Creek in July, 1944. A few weeks later, a huge hog was killed in the same forests, although witnesses said that the creature they had seen was something else.

The biggest post-1909 panic attributed to the Jersey Devil was the fear which gripped Gibbstown in November, 1951. A young boy, Paul May, saw a "half human" creature staring in a window with "blood coming out of its face." This somewhat bizarre detail is reminiscent of the description of a Bigfoot creature seen in northern New Jersey in 1976: "blood was coming out of its eyes," said one young witness. This detail is generally thought to be a reference to the red eyes (whether by the phenomenon of eyeshine or self-luminous varies from witness to witness) often reported on Bigfoot creatures (most notably northern New Jersey's 'Big Red Eye').

A number of hunters set off into the forests around Gibbstown in search of what was taken, once again, to be the Jersey Devil. They heard frightful screams: "It sounded like some birds in the Philadelphia Zoo who are chained, or some other animal in distress, but it was bloodcurdling and eerie," said Ronald Jones.

Another hunter, Jerry Ray, claimed that the monster grabbed at him with "a wild look in its eyes." Soon, the police were nailing up signs proclaiming that *the Jersey Devil is a hoax*, and whether it truly was or not, sightings died down. The general description, though, was of a half-man, half-beast, 7 feet tall, with an ugly face. Some said it was a chunky man with an animal's face. Regardless, it wasn't exactly the typical horse/gargoyle appearance usually attributed to the Devil.

A white-furred, red-eyed monster was seen in the spring of 1975. The animal was about 7 feet tall, and was seen in a wooded area near some apartment buildings on the outskirts of Pennsville. These woods are near a large marshy area and wildlife preserve. The witness also says that he often hears screams and grunting sounds all along the nearby sections of the Delaware River.

Back in Delaware County, though, another sighting possibly of the Wildman of Leiperville was made by Daniel Flynn early on the morning of January 20, 1909. He was on his way to work, walking along the Chester Pike, when he saw a thing emerging from a yard, a thing which "had skin like an alligator, stood on its hind feet and was about six feet tall" and which fled up the road "faster than the speed of an automobile" when it saw him. It is sometimes suggested that the 'scales' of so-called lizardmen may actually be the matted or wet hair of an apish entity, and the possibility is thus raised that this reptilian creature was actually the Ape-Boy, having taken a midnight dip.

And as Flynn's scaly sprinter was seen in the height of the Jersey Devil mania sweeping the Delaware Valley throughout that week in January, it provides a convenient segue into our next chapter.

# Chapter Fourteen
## The Jersey Devil

*Oh the Devil's loose in Bucks and the folks have seen his track,*
*They're keeping up their courage now with jogs of applejack.*
*It lives in jugs and barrels and leaves a hoof-like track,*
*And you can see it plainer when you're full of applejack.*

- *The Bucks County Republican* **(January 22, 1909)**

The previous chapter linked the tales of the Ape-Boy of Chester with that of the Jersey Devil. During the week of January 16-23, 1909, reports of the Jersey Devil were rampant throughout the metropolitan Philadelphia region. Community events were cancelled and schools closed as panic gripped the hearts of the populace of the Delaware Valley. Children and grown adults were loathe to leave their homes for fear the Devil might swoop down and carry them off to its forsaken lair somewhere deep in the Pine Barrens. Hoofprints, much like the famed 'devil's footprints' of Devonshire in England over fifty years before, appeared in the snow. Never mind that Ivan T. Sanderson found evidence later suggesting that the hoofprints were all a part of an elaborate real estate scheme, or that the wildly differing accounts of the Devil's appearance made it far more likely that most of the sightings were more the product of overwrought minds and less of any sort of real animal.

Though it may have been called the *Jersey* Devil, sightings of the creature were reported from across the Delaware, here in Pennsylvania. In fact, the second sighting of what many researchers call 'Phenomenal Week' came from Pennsylvania – from Bristol (Bucks County). At about 2:00 AM on the morning of January 17, a resident of a house that backed onto the Delaware Division Canal, John McOwen, was awakened by the cries of his baby. Once nearer the window, he heard noises

> "...like the scratching of a phonograph before the music begins, and yet it also had something of a whistle to it. You know how the factory whistle sounds? Well, it was something like that. I looked from the window and was astonished to see a large creature standing on the banks of the canal. It looked something like an eagle... and it hopped along the towpath."

Police officer James Sackville, who later became police chief, saw a weird creature on Buckley Street, likewise along the canal. He said that this creature was winged, and hopped like a bird, but also had animal-like features. It had a screaming voice. Sackville pursued the beast, even firing his revolver at it, but the creature was nonplussed and flew away.

Another Bristol resident, postmaster E.W. Minster, saw the Devil flying across the Delaware, presumably after it flew off following Sackville's gunshots.

> "...I heard an eerie, almost supernatural sound from the direction of the river... I looked out upon the Delaware and saw flying diagonally across what appeared to be a large crane, but which was emitting a glow like a fire-fly.
>
> Its head resembled that of a ram, with curled horns, and its long thick neck was thrust forward in flight. It had long thin wings and short legs, the front legs shorter than the hind. Again, it uttered its mournful and awful call – a combination of squawk and a whistle, the beginning very high and piercing and ending very low and hoarse..."

But are these sightings all they seem? Some information I've turned up by combing the pages of the *Bucks County Gazette* may throw these sightings into the realm of politics, pursuit of the almighty dollar, and – yes, skeptics – liquor.

John McOwen, a liquor store owner, was arrested the year before for violation of the Brooks High License Law. That law placed stricter regulations on both the granting of liquor licenses, and on the illegal trade of alcohol. In essence, it was an early attempt to limit the number of alcohol-peddling establishments, a prelude to the full prohibition passed a few years later. On this occasion, his bail was paid by none other than postmaster E.W. Minster; one can assume that Minster would stand to lose out financially if McOwen were jailed (see below). McOwen was later acquitted. McOwen had also been publicly trashed in the press a number of times; Dr. E.J. Groom "did not approve [of] Mr. McOwen's way of doing business;" Henry Rue said McOwen "did not keep very good liquor;" Ray Buseman and John Milligan said the same in the same article.

'Postmaster' Minster also seemed to be far more than postmaster. He was owner of the Artesian Ice Company, still in business today, and just eight days before had received a payout from the Pennsylvania Railroad Company involving an accident in which one of his ice wagons was destroyed and two horses killed at – you guessed it – Mill Street near John McOwen's liquor store. Minster also had assumed the rights to haul the steamer engine used by the Bristol fire department. And the Jersey Devil sounded like a factory whistle…

I'm not certain how Officer Sackville could figure into this, but I do find it somewhat odd that Police Chief Saxton (the chief preceding Sackville) said "that if granted a license he would resign his present office and go into the liquor business extensively, especially supplying malt liquors by the gallon". And in fact, in 1891 Chief Saxton had submitted an application to open a wholesale liquor dealership on Bath Street – the location of McOwen's residence. Appar-

Police attempt to quell the 1951 panic at Gibbstown, New Jersey.

ently, this was rejected, since 20 years later Saxton was still chief.

It was also mentioned that at the time of McOwen's arrest came "just as [liquor store owners] are inaugurating a vigorous political campaign against candidates for the Legislature who are in favor of local option."

I'm not quite certain how these things fit together, but it seems to throw a different light on the whole affair.

\* \* \* \* \*

Nat Thompson, a resident of a house at 20th Street and Edgmont Avenue in Chester (Delaware County), saw around his house a number of tracks. Upon further investigation, he found a rough triangle was formed by the places where the tracks occurred – a triangle bounded by his home, the Enameling Company's buildings at 15th and Earey Streets, and the Chester Rural Cemetery. The tracks looked like those "made by the hoof of a burro... on the front part, claws appear." He had an unusual idea as to the identity of the Devil – "Mr. Thompson has evidence sufficient to prove to any human being that his theory that *the strange thing is a flying machine* is correct" (italics mine). This notion of the identity of the Devil as a machine is an intriguing one – many of the early sightings do, indeed mention a steamy trail and machine-like sounds. Perhaps this notion should be examined more carefully.

Tracks were also discovered by R.J. Knott, living at Edgmont Avenue and 6th Street. He said the tracks led from his hen house, over a fence, to disappear into the Chester Creek. They re-emerged from the creek near William J. Morgan's upholstery shop. The tracks were also found near the Baldwin Locomotive Works in Eddystone.

Regardless, on January 20 (only hours after Daniel Flynn's sighting which closed the last chapter), Mrs. J.H. White went to the backyard of her home on the 1500 block of Ellsworth Street, in Philadelphia, to retrieve laundry. A six-foot tall shape covered with alligator skin stood up and began breathing fire at Mrs. White, who collapsed; her husband came out and swung a clothes prop at the monster. The monster jumped over a fence and into an alley; at just that time a second witness, this time a truck driver on Sixteenth Street, reported a near-collision with the fire-breathing monster.

I've often wondered whether this incident was the basis of the famous "Spring-heeled Jack" episode that supposedly occurred in 1905. In that incident, a series of attacks on women similar to those in London nearly 70 years before were reported; in May, a young laundress named Julia McGlone was attacked by a fire-breathing entity dressed in a crinkled suit as she was leaving the Second Bank of the United States building on Chestnut Street. McGlone screamed, which drew a policeman; the policeman knocked down the creature, which got up, leapt up the stairs, and vanished. Spring-heeled Jack researcher Theo Paijmans has found records of a Julia McGlone in Philadelphia in the latter years of the nineteenth century; incredibly, he found references to attacks, presumably the ones of which the story spoke, but they proved to be of a more mundane nature – a man emerging from the shadows and stabbing women.

Many of the main elements of the White tale are present in the supposed McGlone account, however. Here is the mention of the attacked woman's being a laundress; a similar physical description of her attacker; a male is drawn by her screams; the male assaults the creature; the creature leaps away. I've no real idea how the scene of the attack was moved to Chestnut Street from Ellsworth Street, though; the two streets really aren't anywhere near each other.

Mrs. White's encounter wasn't the only one with the Devil in Philadelphia that day. William Becker claimed to have thrown stones at a nondescript Devil on Limekiln Pike; later that day it was seen again, presumably making its way back across the Delaware, by Martin Burns and several others at the intersection of Fairmount Avenue and Beach Street (that stretch is now Delaware Avenue).

* * * * *

Early on the morning of the 21$^{st}$, William J. Maloney was working the night shift as a guard at Roach's Shipyard (the company closed up shop in 1908 with the death of its founder, and the property sat vacant until 1913) when he heard a series of growls, howls, and screeches, "like a couple cats fighting," in his words.

He followed them to the boiler-house of the shipyards, and looking in saw "a couple of round lurid objects which looked like balls of fire." Maloney fled, and the fire-eyed monster vanished. It should be noted that Maloney denied the story when reporters from the *Chester Times* asked him about it, although whether that was because it was just a rumor or because he just showed off what caliber of security guard he was is unknown.

Thursday night, January 21, the Devil put in an appearance near a lumberyard in Wycombe (Bucks County) and the witnesses said that the creature was "part animal, part bird, part buzzard." A helpful description. The witnesses' descriptions of its size weren't any help, either – some felt it was 9 feet long, while others felt that it was upwards of 20!

A few hours later, it returned to Chester. A respected citizen – no name given – heard noises on top of Robinson's brick-sheds around 10:30 PM.

> "Suddenly, there was a floundering and struggling in the center of Engle street, and out of the confusion there arose a strange-looking animal – half-beast and half-bird – with wings like a bat and a long tail the end of which looked like the point of an arrow. The weather was very foggy, but by the glare of the electric lights, this citizen saw the strange-looking animal fly down Engle street. As it neared the elevated railroad it seemed to rise like a big airship, and passed over the tracks just as a northbound express train was passing.
>
> Whether it was a beast or bird, it was of sufficient size to cause the engineer to sound a sharp danger signal from his whistle. The strange creature was next seen soaring over the top of the borough hall and alighted near the public school building at Third and Jeffrey streets. At this point several of the police officers started in pursuit of the beast or bird, or whatever it is. They gave chase up Commerce street to the rear of the undertaking establishment of Thomas Minshall, where the animal was lost in the fog. The tracks of the creature could plainly be followed in the snow and people living in the neighborhood say that they heard distinctly the rustle of wings and the clatter of strange feet at a late hour.
>
> An examination revealed the fact that several of the telephone wires were torn down and it is supposed that this was done by the animal coming in contact with the cables during its flight."

But the Devil wasn't done putting in appearances that night. As the night of the 21$^{st}$ turned into the wee hours of the morning of the 22$^{nd}$, more poured in.

\* \* \* \* \*

J. Vernon Williams of Middletown (Delaware County) was awakened by barking dogs around 2:00 AM. The dogs were circling around some animal which they seemed loathe to engage. To complete this nondescript encounter, the "strange animal" fled when Williams opened the window.

Later that morning, the Devil was back in Chester, where he frightened two girls by flying out of a boxcar. C.C. Hilk, a Trenton barkeep, was told that the Devil had been captured in the barn of his Morrisville (Bucks County) farm. It seems the creature had been perched on a wagon, and the door slammed and bolted when it was driven into the barn. But of course, the Devil wasn't there when Hilk arrived.

It also appeared near the baseball field at the corner of 11$^{th}$ and Pike Streets, seen by two police officers named Weller and Reber. They thought they had come across a stray horse or some other animal, but as they approached it, it faded from view, leaving behind only hoofprints in the snow. This sounds more like an encounter with some sort of phantasmal animal than the Jersey Devil, though.

And the week of January 16-23, 1909 and all it carried with it are left behind. But the 'Jersey Devil' sightings didn't die off. Far from it, in fact.

Not to be left out of Jersey Devil mania, on January 27$^{th}$ (just a few days too late, sorry, Wilmington), the *Morning News* of Wilmington, Delaware ran a story detailing the sightings of the Jersey Devil in that city during the week, though unfortunately not giving dates to any of the events. Tracks were found, as they were nearly everywhere else – this time at the railroad freight yards at Todd's Cut (near Fort Christina, where the first Swedish settlers landed in 1638). Lewis G. Spence found tracks similar to those Nat Thompson had found in Chester.

Another Thompson had one of the more interesting sightings of the Devil that came out of the city. At about 4:00 AM on one of the days, William Thompson, a milkman (sound familiar? see Bishop's sighting, last chapter) saw what he took to be a calf near Shellpot Park in north Wilmington. As he neared it, it rose onto its hind legs and he saw it had a "long neck, long tail, long nose, short forelegs with claws at the end, and long back legs upon which it was sitting" – much like a red-eyed kangaroo. It barked at him like a dog and Thompson fled in abject terror.

That night, a group of men travelling along a residential street, Concord Avenue, saw something skulking just out of sight. They ran from the creature – as this was right across the street from Shellpot Park, it's likely it was the same animal. But then, as it was in the shadows, it really could have been most anything else. Three firefighters encountered what they thought was the Devil in the vicinity of Clayton Street and Lancaster Avenue. It jumped over the ticket booth of the Clayton Street ballfield – the firefighters said a distance of 25 feet (a little bit of an exaggeration, I'm sure).

It is worth mentioning that the presence of at least some mountain lions in the Wilmington area was confirmed in recent years. A video was taken near Arden around 1995, and one was then seen at White Clay Creek, near Newport, in 2002. Another sighting was investigated at Lum's Pond, just above the Delaware-Chesapeake Canal. One was seen as recently as November 10, 2010, while I was in the midst of writing this book, at Pike Creek.

A creature that some called the Jersey Devil on September 9, 1910 killed a bloodhound near

Springvale, a small village in Windsor Township (York County). William Smuck saw the being, which he declared to be "about the size of a large dog, but with legs shaped like those of a kangaroo." Other witnesses said that the being had quills. In November 2010, another dog in Windsor Township was mysteriously killed, and claw marks found in a screen door in the back. The owner felt it was done by a bear, although some thought the claw marks in the screen more in line with a mountain lion. Other trash cans in the neighborhood had been knocked over and rummaged through.

People at the *National Hotel* in the tiny village of Fallsington (Bucks County; the building is now called the *Stagecoach Tavern*) woke to find a great number of tracks leading round and round, on the morning of September 1, 1911. In a case quite reminiscent of Williams' 1909 encounter, no one actually saw the beast, though the neighborhood dogs were noisy the night before and several bore the signs of having got in a tussle with something. The *Bucks County Gazette* summed up the feeling: "It is thought that some animal may have escaped from the animal farm near Yardley, or else the 'Jersey Devil' has been visiting in this vicinity."

Louis Hirsch of Philadelphia claimed to have captured an infantile Jersey Devil in November, 1911. The pet dealer said that he found the small creature in one of the large goldfish tanks he had in the backyard of his home.

> "The creature is two inches long, with a head like a horse and a tail like a tadpole. On each side of the head are three horns. There are four legs, each with five nails on them. The toes are separated more like fingers, with one that would represent the thumb apart from the other four. The color of the creature is drab, with speckles of darker tone, and the stomach is yellow as gold."

The article goes on to say that the little monster fits the descriptions of the Jersey Devil from 1909 "in every description," which of course is a lie as there was hardly any consistency to the reports and certainly nobody saw anything like this! As to its identity, it's nothing exotic: I would suggest either a small mudpuppy or an axolotl (*Ambystoma mexicanum*), both of which match the animal in coloration. The gold belly (which is not natural) suggests a hoax, though whether the salamander was put into the tank by Hirsch or some other prankster is unknown. A mudpuppy's head is longer and would more likely be called "horselike," though the axolotl fits the size of the animal better.

In July, 1912, a weird creature, said by some to be the Jersey

## The Mystery Animals of Pennsylvania

Devil, was pulled from the Delaware River. The account of the capture, appearing in Iowa's *Sioux County Herald*, is reproduced below in full:

> "Attacked by a mysterious creature, variously described as a "monstrous amphibious animal" and a "furious freak fish," Daniel Miller, a shad fisherman, was rescued from harm by his companion, Harry Taylor, in the Delaware river, off Gloucester [City], N. J.
>
> For five minutes after the creature had been hauled into their boat it gave battle. Until Miller became exhausted he fought the attacks with his fists. Taylor, who was operating the boat, was afraid to leave his position while the battle waged, for fear the boat would capsize.
>
> Miller, who is one of the oldest fishermen in Gloucester, accompanied by Taylor, set out with nets for shad early in the morning. Shortly before noon, when they were preparing to return home with their supply of fish they pulled in the net.
>
> *Leaped at Captor.*
> As they brought it to the surface the weight became noticeably heavy. There was a constant jerking at the ropes. At last they were unable to pull the net any higher and lashed it to the side of the boat. When Miller opened the net the creature sprang from the water at him. The force sent Miller sprawling into the middle of the boat. The creature, which had two rows of long teeth, snapped at him viciously.
>
> When the fishermen finally landed their "catch" on land, hundreds of persons flocked to see the creature. It was taken to Miller's boat house. Fishermen, who have seen and caught many kinds of fish, shook their heads when asked what they thought the thing was. The creature measures five feet, four inches from the head to the end of the tail. The tail alone measures three feet. The head resembles the head of a large "snapper" and is not unlike that of an alligator. It has 20 teeth, some short and ragged, while those in the front are shaped like a dog's, long and pointed.
>
> On its broad gray back are scales from one to two inches in length, which overlap one another. The creature has four feet, like those of an alligator. When standing it would be about a foot from the ground. The tail, shaped like a huge cone, resembles the tail of an alligator.
>
> *May Be "Jersey Devil."*
> A number of the oldest fishermen believed at first that the creature was a lizard more than two centuries old, while others declared that the thing answers to the description of the "Jersey Devil" which several years ago gave the entire [e]astern states a fright.
>
> The thing will be kept by Mr. Miller at his bath house and placed on

> exhibition. In the meantime the scientific authorities at Washington have been asked to examine the creature and pass judgment as to what it may be."

I'm going to make a wild leap of logic here. Since they make the point no less than three times that this thing had features resembling an alligator, could it have been… an alligator? (Probably, more likely a caiman, as these display more noticeable "overlapping" of scales on their back).

January of 1919 saw yet another sighting of what was called a Jersey Devil by some. It seemed that a white figure was seen wandering the lanes of Rockdale (Delaware County) by night; how this figure qualifies as a Jersey Devil, I'll never understand.

\* \* \* \* \*

In February of 1924, two men, A. Parke Patrick and Herbert Andress, were driving along the road north of West Chester (Chester County) when they saw a large animal amble out into the path of their car. The animal, which the men described as "a raccoon of enormous size," was callously attacked with a starting crank (the two men would have been in an old-style hand-cranked car, remember) by Mr. Andress, and then pursued him back to the automobile. Once there, it ran alongside the car until the men reached a golf course a half mile away. Several times, the animal attempted to make its way into the car, clambering onto the runningboards of the auto. The sighting was followed a few years later by another of a creature, that came to be dubbed the "Dorlan Devil," from nearby Lyndell.

On January 21, 1932 both the *Coatesville Record* and *West Chester Daily Local News* described a monstrous creature encountered in the woods by John McCandless. I had become aware of the McCandless encounter some years previous, when as an impressionable young lad I read Loren Coleman's *Mysterious America* for the first time. Coleman's book had a brief paragraph describing the sighting, and since the time I first read it I was well-nigh obsessed with tracking down the story. I finally tracked down the original references only a few months before sitting down to write these lines.

It seems that on January 19, McCandless was directing his crew of lumberjacks when he and another man named Lee Yeager went to investigate a peculiar groaning sound coming from near their worksite. The monster sprang out of the bushes towards the men, who promptly turned tail and ran, with the creature in pursuit, to retrieve a shotgun from the home of an old farmer in the area. The farmer concurred that he, too, had seen the creature one night a few weeks before. When McCandless and the workman returned, gun in hand, they found some large footprints in the mud but no creature.

> "It was about the height of a man and was without clothing of any kind. Its skin was a yellowish-gray in color and its face was more horrible than that of any animal I have ever seen pictured. Its head, hands and feet appeared to be unusually large and it stood partially erect when walking. I could not say whether it was man or beast."

**The Second National Bank building, where Julia McGlone was supposedly attacked by a Springheeled Jack entity in 1905.**

The workman who encountered the creature with McCandless wasn't so ambivalent when it came to the human or animal question.

> "Man, that wasn't nothin' human! It growled and groaned like a wildcat and made chills run up and down your back just to hear it. It didn't walk either; it crawled and crawled mighty fast.

The *Chester Times*, which had reported many Jersey Devil encounters in 1909, took a somewhat more skeptical view of the encounter. That newspaper said the beast was "believed to be a maniac," "a Jersey madman," and also provided a different quote from McCandless which provided some additional information:

> "I couldn't get a clear glimpse of the body, but I know that it was not hairy, like an animal's body would have been. I saw the hands very clearly, however, and they were just like human hands, only unusually large."

The creature was believed to be a deer by many residents of the area. But after a few days, on January 22, McCandless, his lumberjacks, and a local man named Josiah Hoopes combed the region in a futile hunt for the beast. A group of four local reporters armed themselves and

ventured into the forests of Lyndell "to personally interview his Satanic majesty the Jersey Devil or the member of his royal household said to inhabit that particular section of Chester County." Apparently, Satan was feeling rather antisocial that day, because the reporters didn't find anything.

In July, 1937 the newspapers were once again to carry news of the monster. This time it had a slightly different description; on July 28th, Cydney Ladley from Milford Mills (that town was inundated with the 1972 creation of the Marsh Creek Lake), Mrs. Ladley, and a neighbor named Mrs. Smith were driving towards Dorlan when a creature leapt from the trees.

> "It leaped across the road in front of my car. It was about the size of a kangaroo, was covered with hair four inches long and it hopped like a kangaroo. And eyes! What eyes!"

Inevitably, a posse of armed farmers combed the woods and marshes around Milford Mills, but nothing turned up.

Interestingly, the Shamona Creek flows into the Brandywine Creek near the intersection of Shelmire and Dowlin Forge Roads just south of Dorlan. The Lenape word *shamong*, as I have noted earlier, means 'the place of the horn' and has been associated with white ghost stags and the Jersey Devil.

Two years after Ladley's sighting, another creature surfaced in Chester County. This one appeared in Westwood, a small residential area located west of Coatesville. A 31-year old man named Sylvester Scott was spreading fertilizer on February 15, 1939 when he saw the creature that had been making some horrid nocturnal screeches "like that of two cats" (what?) over the last few weeks.

> "It stood two or two and a half feet off the ground. It was colored like a deer in front, with white on the flanks. It had paws, remember, and not hoofs [sic].

It jumped just like a deer, about two feet up in the air. It had a very small head on the end of its neck, which was a foot long anyway. It was a fast runner, all right. It ran away from my dogs – beagles – and they didn't seem to want to follow. Neither did I."

People theorized that it was a chamois (*Rupicapra rupicapra*), or possibly a springbok (*Antidorcas marsupialis*). A week later, though, some unnamed 'authority' deemed it to have been a 'roof rabbit.' Roof rabbit was a colloquial name for a big cat, it seems.

We'll leave this chapter, though, with the most infamous "sighting" of the Jersey Devil in Pennsylvania, at a museum at $9^{th}$ and Arch Streets in Philadelphia. J.F. Hope provided a tale of how an armed posse of men moved through the city in pursuit of the creature, eventually capturing the Devil after pelting it with snowballs (?). Bloodied, the men dragged the "Australian Vampire" back to the museum. In actuality? A handful of men hired from a local tavern had chased down and captured a kangaroo, which Hope and Norman Jefferies had painted green, attached wings to, and put in a basement room with dramatically-gnawed bones. Which, of course, the kangaroo couldn't have cared less about.

# Chapter Fifteen
## Montie the Monster

I had read a tantalizingly-short account of the creature which appeared near Pottstown in 1945 years ago, and I can still remember the picture that accompanied the short write-up – a gun-clutching farmer in the foreground, a semi-humanoid creature leaping away – vividly. It was one of those stories I pursued for years. It was the tale of the Sheep's Hill Thing, otherwise known as Montie the Monster.

Several witnesses reported a leaping, wailing creature from Sheep Hill, in North Coventry Township (Chester County) not too far from South Pottstown. John Hipple claimed to have seen the beast in Montgomery County. "It was like a big cat," he said. "I shot at it and it leaped 20 feet into [the] air and, screaming, disappeared." Another witness was John Wojack, who said that "it gave a shrill cry, then it bounded away in leaps of at least 10 feet in length each." Lester Thompson gave a harrowing account of the noise it made – "it starts sort of low-pitched, gives a couple of short bursts first then lets her rip." A young boy said it sounded like a man "screaming as loud as he could." Shades of the Westwood monster's yowling. The media summed up the reports of the monster:

> "The mysterious Montie which caused the furore has been variously described as a panther, a chow dog, and a puma. People who swear they saw Montie say he is about three feet long and has a bushy tail 'with a kink in it.' Some claim the animal has the wail of a banshee, others say he barks like a dog, while still others insist he laughs like a hyena."

That may be the corpus of sightings – or, mostly hearings – of the monster, but at this point hysteria set in. Gangs of hunters roamed the woods and fields of Sheep Hill, searching for 'Montie the Monster.' One resident complained of the hunters, "they are not apt to shoot the wolf or whatever it is, but they will likely shoot me if I am a foot above the ground." It was said that it was "worse than the battle of Gettysburg" on Sheep Hill. One resident even declared that she was nearly ready to start opening fire herself.

The headline of one media report summed the situation up aptly – *TRIGGER-HAPPY HUNTERS ARE MORE DANGEROUS THAN MONSTER MONTIE.* No less than four people were injured by other hunters – two were shot, and two were injured in a car accident when the driver thought he saw Montie. Following these accidents, the state police began fining any would-be monster hunters found prowling the woods. Game Warden Peter Filkowsky placed bear traps in the woods, but two days after the hunt, he had failed to capture anything. In November, 1948, there were further sightings of what may or may not have been Montie. He or she was wandering closer to the Schuylkill River and closer to Pottstown now. Montie II, or so we'll call this beast, was described as "a black cat-like animal about the size of a bird-dog. It leaps across the fields with a bounding motion more than it runs…" But no more was heard of the 1948 beast. The fact that both versions of Montie visited Chester County in November of their respective years is surely significant.

In 1973, another mystery beast of similarly Montie-like habits was described only as having piercing red eyes and a sulfurous smell. Adams describes the 1973 being as winged and cat-like with large claws. As described in *Varmints*, though, the Pottstown articles on the beast haven't yet been tracked down. An article in the *Doylestown Daily Intelligencer* (March 16, 1973) notes that a large rabbit hutch was tipped over and smashed by some unknown animal or person in Bucks County, and said article makes note of investigations taking place in the Pottstown area, but until the relevant articles in the *Mercury* can be tracked down, this will have to remain a mystery.

The self-same area of northern Chester County and southern Berks County between the Schuylkill River and Chestnut Hill was home in the 1840s to a bizarre religious cult led by Theophilus R. Gates, originally from Connecticut. They called themselves the Battle-Axes of the Lord, and they settled in a place they called Free Love Valley (below). Gates' sect took a

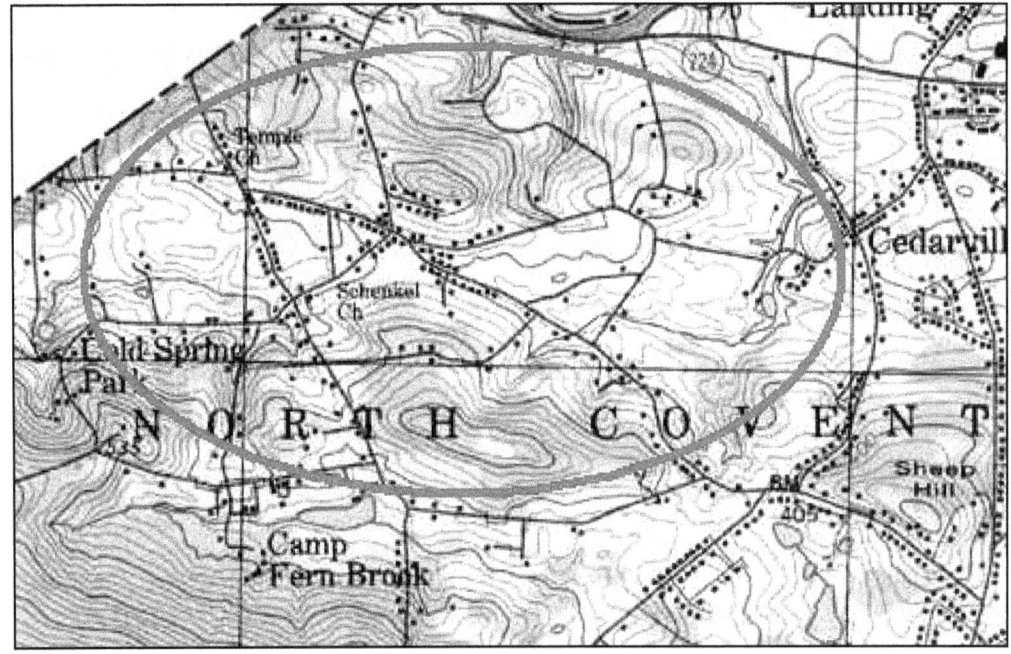

stand against religious intolerance and Federal government. The Battle-Axes would gather at a place they called the Temple, "clothing and morals abandoned in the ecstasies of being perfect". They also engaged in a protest, walking stark naked down the aisle, "in their shamelessness," at the Shenkel Church after its minister condemned the Battle-Axes. Gates was apparently "touched" – an incident is recorded in which he attached shingles to his arms and leapt from a roof, attempting to fly to Heaven. But after Gates' death, the cult was to die a quick death, and was dissolved by 1857.

The area is also home to another sordid piece of history, the murder of Hannah Shingle and its aftermath. Hannah Shingle was an old reclusive crone who lived in the valley in an area called Hell's Hinges. On October 21, 1855, a handyman named John Miller discovered her lying in an upstairs room, dead, killed by the very hatchet she kept for protection. During the investigation both Miller and many of the Battle-Axes were questioned. A canal worker living in Unionville (Berks County) was found to have blood-soaked clothing but was not charged – you have to love that 1850s police work! Another man, apparently a suspect in the murder, committed suicide by hanging himself – the tree he used as a gallows is now located at the corner of Catfish Lane and Valley View Road.

A man named Montgomery Campbell saw a headless ghost roaming near the springhouse on Hannah's property in September of 1879, and another woman in more modern times claims to have seen a white phantom flitting among the trees by the stream. It is claimed that the Shingle springhouse was used by the Battle-Axes for their orgiastic rites.

As discussed in an earlier chapter, Nick Redfern and others have examined the case of the British "Man-Monkey" which, despite its beastly form, was thought by some to be the phantom of a man who drowned himself in the canal. Could "Montie" have been the transformed spirit of the man who committed suicide in Free Love Valley? The corner where the deed was done is less than a mile from Montie's stomping grounds at Sheep's Hill.

Another possibility must be mentioned. The hill was at the time home to the broadcasting tower of radio station WLAB-FM; the radio station left in 1953, though the tower remains, used now by a television station.

Another tale which could originate from the same area was sent to me by Robert Schneck. He theorized it could have been some newspaperman's response to the extensive media coverage that Fedor Jeftichew, the noted hypertrichotic labelled 'Jo-Jo the Dogfaced Boy' by showman P.T. Barnum, was receiving at the time.

> "HER AWFUL SECRET.
> *An Infantile Monstrosity Taken to an Asylum*
> *And the Parent Jailed.*
>
> A young woman, about twenty-five years of age, named Margaret McGourky, was placed in jail at West Chester yesterday. The facts leading to her arrest have caused no end of gossip in social affairs. She lived all alone in a house in the village of "Hell's Defiance," in the

northern section of the county. She was pretty, and one strange fact about her was that she would never take any of her neighbors further than the front room down stairs, and would but seldom absent herself from home. This was only looked upon as a little peculiarity of hers. She made her living by doing plain sewing for her neighbors.

Several days ago, however, she left the house for a short time. During her absence a strange creature, part human and part animal, was seen to emanate from the house and make for the woods not far away. It ran on all fours and attracted the attention of several men who started in pursuit, captured it after considerable difficulty and returned it to the house. They were horrified to find that its head and face were those of a human being, and the body that of an animal, being covered with short hair. When the woman came back she appeared greatly alarmed because her secret had been discovered. An investigation was made, when it was discovered that she had given birth to the strange being five years ago; that she had kept it in an upstairs room all this time, and that it was about eighteen inches high when resting on all fours. It is a male, and its natural position is that similar to an animal. Its face is very pretty and it has an intelligent look. It is unable to talk, and the only sound it utters are those of a dog. It was taken to the county asylum, where the authorities will take care of it, while the woman was conveyed to jail for cruelty to it."

Get it? A story about a Boy-faced Dog rather than a Dog-faced Boy. But could it have been based on a real event? Newspapers often carried tales of feral or confined children, which were usually ill or mentally deficient. The 'county asylum' to which the child was taken would be the Chester County Poorhouse, the mental asylum not opening until after 1900. There is record of a Samuel Mann, age 6, admitted to the poorhouse in 1886.

But I wonder whether Hell's *Defiance*, mentioned as the location of the story, could have actually been Hell's *Hinges*.

Kangaroos (well, actually wallabies) and alligators (dubbed "crazy crocs" by cryptozoologist Loren Coleman and discussed in a previous chapter) are two of the most commonly reported species reported from far outside of their normal range. Pennsylvania's weird fauna has this covered – in January, 2007 a wallaby (genus *Macropus*, meaning "big foot" – an interesting coincidence in a cryptozoological context) was seen hopping through a backyard in Fleetwood (Berks County), in the Oley Hills about ten miles northeast of Reading. Other sightings of the wallaby began to trickle in to the Berks County Humane Society in the coming days. The nearest zoo, the Lehigh Valley Zoo in Schnecksville (Lehigh County) was not missing a wallaby.

Perhaps, since it was infamously claimed in Australia in 2009 that opium-addled wallabies "chasing the dragon" round and round in circles were responsible for appearances of crop circles, there's a dizzy marsupial behind behind the mystery of the *Hexendanz* (a field in which crop circles regularly appear) in Berks County!

Two wallabies (a male and a female) escaped from a farm in Upper Dublin, north of Ambler (Montgomery County), in 2001. The true culprits behind sightings of giant rats roaming near Ambler, the wallabies were never captured. It is tempting to wonder whether they could have bred in the wild and possibly led to a feral population of marsupials.

Another tale which was resolved happily was that of Melbourne, a wallaby who escaped from a farm in Wrightstown (Bucks County) on June 12, 2007. Three weeks later, Melbourne was cornered at a horse farm in Upper Makefield Township, but escaped; the next day, he was wrestled to the ground at the same farm and put in a horse stall by Roy Silcox and Mike Ragnoli. The owner of the farm from which the wallaby escaped, Charis Matey, felt that the escape was a good thing. "It's almost like he had a vacation," she said of Melbourne.

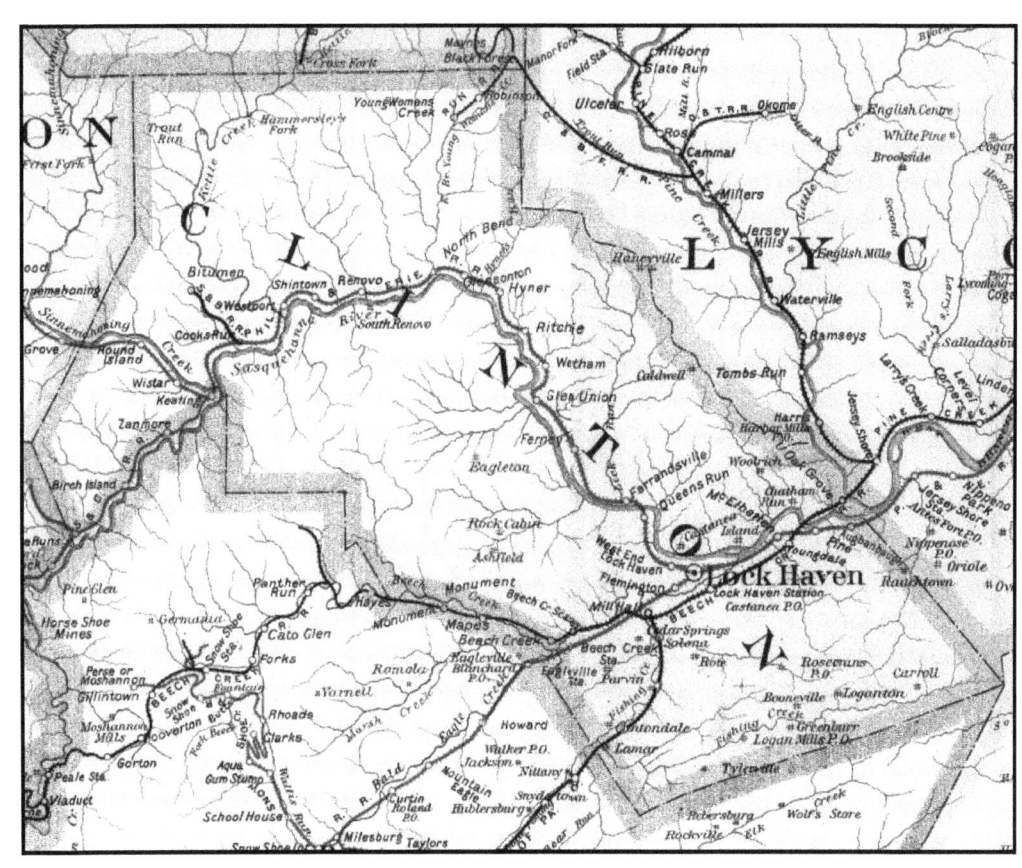

Clinton, PA 1895

# Chapter Sixteen
*Thunderbirds of the Black Forest*

Clinton County is in the north-central part of the state. The county is home to 37,000 people, most of whom are inhabitants of a broad lowland plain approximately 5 miles wide. Heading north from the college town of Lock Haven one finds nearly 600 square miles of land, nearly all part of the Sproul State Forest. A handful of villages and small towns are clustered alongside the banks of the Susquehanna River, but forests and mountains dominate much of the landscape, that traversed by innumerable streams and creeks. The valleys are here and there dotted with boggy areas, the remnants of the Great Swamp, which in times long since covered most of north-eastern Pennsylvania. In short, Clinton County is just the sort of area one would think to be inhabited by monsters, and, thanks to the efforts of a handful of researchers, it doesn't disappoint.

It is in this region that the majority of reports of giant birds in Pennsylvania originate. Interestingly, nearly all are from an interconnected web of runs, creeks and streams, all of which are part of the drainage area of the West Branch of the Susquehanna River. One can even define a fairly clear range if one assumes that these are biological entities. The vast majority of reports seem to be confined to the area of Clinton County described above, ranging eastward into Lycoming County (the reports surfacing from Lycoming are of more recent vintage, possibly suggesting a furthering of the animals' range within the last 40 years). There are a handful of reports from Potter County (directly on the New York border) and it seems that here the birds may have entered the network of streams feeding into the Allegheny River. The number of sightings from the Allegheny drainage area are fewer in number.

Some items of regional folklore concerning the thunderbird are traditions referring to a sighting of a huge black bird by author Mark Twain at Ravensburg State Park near Jersey Shore (Lycoming County) and also to an "exceptional raven" seen by Edgar Allan Poe at what is now Poe Valley State Park (Mifflin County). Folklore would have one believe that this is what inspired his famous poem "The Raven."

Traditions of such monstrous birds, with any of a hundred names but usually called Thunderbirds, exist among nearly all of the Native American tribes and in the traditions of peoples the world over. The Lenape tribe, which inhabited the entire length of the Delaware River (and thus are sometimes named the Delawares) from the state of the same name through Philadelphia and Bucks County and north into the Pocono Mountains and thence into New York state, had traditions of the Thunderbirds. They were variously known as *pléthoak, pëlésëwak* or *pethaxuwéyak* and usually took the form of eagles, grouse, or turkeys. Thunderbirds were mostly benevolent, though several Lenape legends speak of a predatory and easily angered nature. Thunderbird nature or magical power could also be possessed by people, and the companion of the hero Wehixamukes was a "thunder boy".

## Sightings in Potter County

Elvira Ellis Coats, daughter of Richard Ellis, the founder of Ellisburg, was the first to testify to the existence of these monstrous flying creatures in the 1840s. She said they looked similar to vultures, with expansive wingspans. Author Robert R. Lyman saw a brownish bird 3 to 4 feet tall sitting in the middle of Sheldon Hollow Road near Coudersport (Potter County) one day in 1940. The bird had a short neck, short legs, and had a wingspan of 20 to 25 feet. Lyman noted that it rose nearly straight up when it took flight and that it navigated through the trees very well. He also noted that it had narrow wings, an adaptation to help minimize windshear found among water-going birds.

Helen Erway left her school bus near Ole Bull in August 1945. While walking along the dirt road to her home, she saw a huge bird. "It was not an eagle," she said. "The wings were straight out. It made a high pitch noise. The shadow on the road was about 30 feet". Helen suffered mental effects from her encounter: nightmares, insomnia, and an inability to eat (many accounts of encounters with the paranormal or unnatural contain accounts of similar ill-effects on witnesses and this may indicate a more supernatural explanation for Erway's encounter).

Her paternal grandmother, Marion Erway, was a full-blooded Indian and said that she had seen a Thunderbird which was protecting her. She also mentioned that a Thunderbird hovered over Delbert Schoonover, who was bitten by a snake near a dam, until help arrived. The bird was seen by the workers who arrived to help him. I'm not certain whether this is a reference to the 1969 encounter at Alvin R. Bush Dam discussed in a bit.

I'm not sure where is meant by Ole Bull, as no town of that name exists. It is possible the account refers to Oleona, or to the general region of the Ole Bull State Park. Ole Bull was a Norwegian violinist who who first came to the United States in the 1850s and felt that the landscape of Potter County was much like his native Norway. In 1852, he endeavored "to found a New Norway, consecrated to liberty, baptized with independence, and protected by the Union's mighty flag."

He laid plans for several towns – Oleona, Valhalla, and New Bergen (now Carter Camp) – and the ruins of Bull's "castle," Nordjenskald, are on a massive hill overlooking the Kettle Creek. But alas, the colony was not to be, and most of the settlers moved west to Wisconsin.

One of his descendants, Inez Bull, was to encounter a Thunderbird in 1973 near Sunderlinville.

## Sightings in northern Clinton County

H.M. Cranmer, of Hammersley Fork, wrote to *Fate* in September of 1963 about sightings of the birds in that area.

> "I first saw a thunderbird in April 1922. I was standing by my gate at dusk when one flew over, heading north. It passed a pine tree with branches spreading 50 feet, so that I could estimate its wingspread fairly accurately. It was 35 feet. I was alone at the time and never mentioned the sighting for 35 years.
>
> On 27 March 1957, a young man rushed in and said: 'Something big is in the sky'. I went outside and saw a huge bird flying lazily 500 feet above. Its wing motion reminded me of a blue heron, but the bird was lighter and grayer in color. I called the American Legion in Renovo, and inquired if anyone had seen a big bird half an hour before. A man who had just come in said it had flown over Westport, came down Fish Dam Run, and then up Two-Mile Run. He said its wingspread was 25 or 30 feet.
>
> A week later another flew the same course. Two weeks later another flew up Shintown Run, down Hevner Run, and up Hammersley Fork stream. Each time there were several other witnesses.
>
> On 4 July 1962, I saw one flying south, up Hevner Run and down Shintown Run. I called the Legion and told them a thunderbird would come down Shintown Run at 9:15. It did, and flew up Halls Run [opposite Renovo] after crossing over the [Susquehanna] river."

In May, 1964 Charlie Passell and some other men saw a strange bird perched in a tree near the Alvin R. Bush Dam on Kettle Creek. The bird was "larger than a buzzard, real large" and was "definitely no eagle". Passell said that it had a "longer neck than a hawk's but not as long as a stork". Another sighting at the dam took place in the summer of 1969 when Albert Schoonover, engaged in road work with two others, saw a monster bird, whose size is not specified, swoop down and carry away a fawn whose weight they estimated at 15 pounds.

Robert R. Lyman records that in about 1967, at a cabin near the intersection of Robbins Run Road and Carrier Road, a man and wife (who had been seeing huge birds in that vicinity since 1963) saw a massive bird dive at their car. The bird swooped down and grabbed a dead groundhog. The bird seemed to have thin legs and massively taloned feet.

In September of 1972, A.E. Williams and Marvin Wagner were hunting in the area when they saw a monstrous bird sitting in a tree. In 1977, Terry McCormick saw what he at first took to be an airplane until it flapped its wings and flew off over the forest south of Renovo. The bird McCormick saw was grayish brown with a large beak and straight wings and was about the size of a Piper Cub airplane.

Todd Hackenberg and Erin Goundie encountered a bird along Young Woman's Creek near North Bend – near a dam "stood an awfully big, whitish bird. About five feet tall, had a long beak, thick legs as long as the stocky body. The neck was long." The bird had a wingspan of approximately 15 feet and it held its wings straight as it took off.

## Sightings in Lycoming County

Sightings have surfaced since that time along Tombs Run, a tributary of the Pine Creek. The birds were described as being brown or gray with white-tipped wings . The Tombs Run sightings included a few of multiple individuals, though rarely more than two. The sheriff of Clinton County in the late 1960s, John Boyle, had a cabin along the Little Pine Creek, another Pine Creek tributary. His wife saw a monster bird sitting in the creek, which was running low at the time. The bird was a grayish color. Mrs. Boyle estimated the wingspan of the bird at a whopping 75 feet.

There was a flap of sightings of the birds from the Jersey Shore area in the 1970s. On October 28, 1970 Judith Dingler was driving near the town of Avis (Clinton County) when she saw a "gigantic winged creature soaring towards Jersey Shore. It was dark colored and its wingspread was almost like an airplane." Another sighting took place on November 9, 1970 when Clyde and Anna Mincer reported that a monstrous bird had soared over their home.

> "If I recall correctly, it was the 9$^{th}$ and 10$^{th}$ of November 1970 when my wife and I saw the first huge bird which I said had a twenty-two foot wingspread, and I have not changed my mind since. My wife called to me as I was painting spouting and said to me, "Look at the funny airplane." It was riding a jet stream. While I watched it got off the stream, flopped its wings a few times very slow, and was back on the jet stream again.
>
> This bird was right over my house."
> (Letter from Clyde Mincer to Mark A. Hall, Sept. 19, 1977)

On June 8, 1971, Linda L. Edwards and Debbie Kraft saw a bird, whose wingspan was at least 18 feet, feeding on a dead opossum on Cement Hollow Road. The bird seemed to have launched itself straight upwards into the air rather than running to take wing. "It just flapped its wings twice and was off', Mrs. Edwards said. During his 1940 sighting, Robert R. Lyman was also amazed at the maneuverability of the bird. The bird was seen along Larry's Creek by Clair E. Koons and Wilson Frederick on August 7.

On September 23, 1972, Mary and George Missimer saw a dark bird riding air currents and appearing to be as large as a Piper Cub airplane. The bird was seen again on the 25$^{th}$ flying towards the Renovo area of Clinton County discussed earlier. When he was visiting Mary on the 28$^{th}$ of that month, Clyde Mincer claimed that a huge bird was seen flying towards Mount Logan in Clinton County.

An undated account, but one presumably from the early 1970s, comes from Dale Williams, who was hunting with Harry and Bill Pepperman around Salladasburg when they saw a gigantic bird. As of 1993 the birds were still in the area, because in July of that year Shane Fisher and his family saw a bird at their home along Larry's Creek. "You could hear the air rush through the wings", Shane said. The bird was similar to an eagle with large black eyes and was "wider than the cross bar on a telephone pole". Gerald Musinsky adds that the bird seen by the Fishers was similar to a Steller's sea eagle, a description whose significance will soon become apparent.

In March 1973 Mr. and Mrs. Joseph Kaye of Lock Haven were driving near the Oregon Hill Ski Area in the northern part of the county along the Little Pine Creek (site of the Boyle sightings). As Mrs. Kaye later told Mark A. Hall,

> "We saw a large object ahead of us on the right berm of the road. At first – as we drove toward it – we couldn't identify it but both my husband and I saw it. Then it started toward us and we realized it was a huge bird. Its wings were so big it had trouble getting off the ground. The wings flapped slowly and heavily ... The one wing swept our windshield as it rose very slowly. I could see white feathers on the head and on the feet – body appeared to be black – rather like a bald eagle – as it came even with my side car window. It seemed too heavy to get off the ground. Then we were past it and it disappeared over the trees. We were so awe-struck at the whole thing that we could hardly speak."

In November 1989 Shannon Breiner was at her grandmother's home in Barbours (Lycoming County) when she saw a "large eagle-like bird" with a short, thick neck and broad chest rise onto two legs and make for a swampy area. The bird seemed to have a nest built of grass at its feet. Shannon was adamant about the fact that the bird was "not a frail bird" like a stork or heron.

Very near to the location of Shannon's sighting is Hughesville, where one evening in April (the account is otherwise undated) Tammy Golder encountered a similar bird while driving Beaver Lake Road over Frantz Hill. She was driving at about 40 miles per hour when a bird neither turkey nor vulture struck her car and left a dent.

Kim Foley was driving with her son near the Mt. Zion Church cemetery near Montgomery (Lycoming County) in September 1992. She saw a dark brown to black bird eating a dead deer. She likened the bird to a photo of Steller's sea eagle (*Haliaeetus pelagicus*), although this bird was larger. Only two weeks after Foley's encounter, Dave Sims saw a similar bird near Hyner (Clinton County). It was dark gray and flew at least 55 miles per hour.

## Sightings in other counties

In 1892, a Tioga County man named Fred Murray was at a lumbermen's camp at Dent's Run (Elk County). Murray saw a flock of birds resembling buzzards (note that the word 'buzzard' is colloquial word referring to a number of species of vulture, and does not refer to raptors of genus *Buteo*), but with wingspans of 16 feet or more. The birds must have stopped at Dent's Run for a few days, because an ornithologist came from Pittsburgh and observed the birds, as well.

That same year, a Centerville (Crawford County) farmer captured a bird which had been scavenging a dead cow on his property. A.P. Akeley saw the bird in captivity, which he said was between 4 and 6 feet tall and gray.

In the summer of 1973, Mike Floryshak saw a bird gliding above his car as he was driving across farmland near Huntsville (Luzerne County). "The wings appeared to be twenty feet wide", said Floryshak. It was a brownish color, with a large beak and stiff wings.

One area in which the birds are reported often is around Curwensville (Clearfield County). Debbie Wright and Sue Howell saw a bird which was larger than their car near Drocker's Woods one morning in the spring of 1977. "It was big, black or very dark brown with a huge beak," Wright said of the encounter. Truckers reported that they also had been seeing "big buzzards", dark in color, on Boot Jack Hill near the Wright/Howell sighting for years. And on two mornings in the autumn of 1991 and October, 1992, a Curwensville woman saw a large bird standing in the Susquehanna River on one leg with its head tucked under its wing. It had thick legs, looked similar to an ostrich, and wasn't a heron.

In June of 2002, two of the birds were seen near Tunkhannock (Wyoming County) and were described as being similar to a crane or heron, but larger and with shorter legs. The size of the birds was somewhat of a mystery – "wider than the span of the branches", writes Matt Lake in *Weird Pennsylvania*. The witnesses felt the birds most resembled the little black cormorant (normally native to Australia), but far larger.

Russ Powers and Denny Eckley saw a bird attack a yearling fawn near Bear Run (Tioga County) in the Algerine Swamp. The bird had a longer beak than an eagle, dark grayish brown plumage, thick legs, talons larger than a man's hand. It had downy feathers on its head and a wingspan of between 14 and 16 feet. Later, the two claimed to have seen the infamous "Thunderbird photo" in the *Guinness Book of World Records* (this photo depicts a huge bird nailed wings outstretched to a barn door, flanked by gun-toting cowboys and was supposedly taken in Tombstone, Arizona; however, the photo has never turned up although Ivan T. Sanderson claimed to have seen it and H.M. Cranmer claimed to have it).

## Musings

It is possible, if not probable, that some of the place names in the relevant areas may be inspired by the birds as they have been a part of regional folklore for two centuries now. The town of Avis (Latin for 'bird'), Ravensburg State Park and a mountain called the Ravensberg, and Crane Island are both located very near Jersey Shore which as previously discussed has been a hotspot for sightings of the monstrous birds. It is also possible that the name Phoenix Run, a tributary of the Pine Creek which flows past Sunderlinville (Potter County) where Inez Bull and her mother encountered a Thunderbird in 1973, is a reference to the birds. Carrier

Road, where the late 60s encounter described above took place, also seems to reference a predatory bird.

Genesee County in New York (near Batavia) is a mainly pastoral landscape. In the late 1980s, a skeletal bird was discovered here which was found to be of *Gymnogyps californianus*, the California condor. Theories about an eastern-ranging population of condor had circulated for a time in reference to the mystery birds in Pennsylvania. Both author Robert R. Lyman and H.M. Cranmer had theories about the eastern condor they termed *Gymnogyps pennsylvanianus* and the finds in New York seem to support those ideas. In his article "Return of the Thunderbird," Clinton County native Gerald Musinsky notes that while one might presume that such a bird of the size given in most thunderbird reports could not possibly live in the dense forests of northern Pennsylvania, the two largest species of eagles in the world, the harpy eagle (*Harpia harpyja*) and the monkey-eating eagle (*Pithecophaga jefferyi*) live in thickly forested terrain along rivers – and the reports bear witness that thunderbirds, as well, seem to frequent waterways.

Kim Foley likened the creature she saw in 1992 to a Steller's sea eagle (*Haliaeetus pelagicus*) as did Shane Fisher in regards to his 1993 sighting. This is one of the largest eagles. Native to the coastal regions of the Kamchatka peninsula in Russia, the Korean peninsula, and the northern coast of the island of Hokkaido in Japan, the sea-eagle feeds mainly on fish. Interestingly, the bald eagle (*H. leucocephalus*) widespread throughout much of the United States and the north-central part of Pennsylvania in particular, is also a member of genus *Haliaeetus* and has similar habits. Could a mutation be causing some bald eagles to appear similar to this Asian relative?

The islands of New Zealand were once home to a truly monstrous predator called Haast's eagle (*Harpagornis moorei*). Thought to have become extinct in the 1400s, it inhabited the plains and forests of the islands, and is thought to be the basis for the beings known variously as *pouakai*, *hokioi* or *hakawai* in the legends of the indigenous Maori. These were man-eating birds of immense size, in some legends thought to be air spirits almost similar to the Native American conceptions of the Thunderbird. The Haast's eagle is believed to have taken wing aided by a leap propelling it upward and then a use of extremely strong wing muscles. It's fairly likely that the thick legs mentioned often by witnesses to these monster birds is a hint to just such a method of taking flight among them; the mention sometimes made of birds taking off "straight up" or "like a helicopter" could be a reference to this taking place.

The wingspan of the Haast's eagle appears to have been roughly 9-10 feet, greater than the massive Harpy eagle which is the largest currently-extant species of predatory bird. But even the massive size of a Haast's eagle, which was every bit capable of preying on the huge moa of ancient New Zealand and, most likely, on humans is dwarfed by even the average estimates of size for Thunderbirds. In fact, the only bird of that size is the now-extinct teratorn (several species of prehistoric condor and vulture relatives), specifically *Argentavis magnificens*. *Argentavis* had a wingspan of anywhere between 20-26 feet and stood five feet tall when standing. Bernard Heuvelmans and others have noted, though, that an accurate estimate of the size of a bird in flight is extremely difficult to achieve and thus the 15-17 foot range usually given may be exaggerating the actual size of the bird. Karl Shuker and others feel that the golden eagle (*Aquila* c*hrysaetos*) may account for some of the reports. The golden eagle is an impressive bird, smaller than the bald eagle; its normal range is in the Rocky Mountain regions of the West and further north in New England and maritime Canada. Although they do range into Pennsylvania during the winter months, it would most likely be an out-of-place and unfamiliar animal to most residents of the region and is sufficiently different in appearance (at least coloration) from the bald eagle. Indeed, the thunderbird seen by Alison Stearn near Shingletown (Centre County) was likened by her to a golden eagle.

## Thunderbirds and Mothman

Researcher Mark A. Hall has analyzed the dates of the various sightings and found that many of the Pennsylvania-based sightings seem to originate over the spring and summer months, with sightings originating from areas to the south during the autumn and winter – including the sightings of West Virginia's monstrous Mothman in November and December, 1966.

In regards to Mothman, John Keel in his book on the subject, *The Mothman Prophecies*, notes that there was a sighting of some sort of unusual avian in western Pennsylvania during the timeframe of the Mothman affair. On November 26, 1966, hunter George Wolfe, Jr. saw an unusual bird in a field near Enon Valley, in Lawrence County (not in Beaver Falls as Keel wrote. Wolfe was from that city, however, which may have accounted for the confusion). It was nearly seven feet tall, and Keel notes that Wolfe said the bird rocked back and forth as it moved. It looked something like a gray ostrich, with "a long neck and a round body with a plumed tail that reached high above its body" – a description which, to me, sounds more like a large turkey than an ostrich.

But that may not have been the case, as only four days later on November 30, 17-year old Dave Hoffman reported that he encountered "the biggest [bird] I ever saw" along the Big Sewickley Creek, on the border of Allegheny and Beaver Counties about 20 miles south of Wolfe's sighting. The bird was light-colored and over six feet tall – newspaper reports speculated as to whether it was the same bird seen in Lawrence County. Hoffman said that this bird, though, "looked like a crane – an awful big one. It had long legs and appeared to have something wrong with its wings when it tried to fly." In fact, it crash-landed in the woods when it tried to.

John Keel noted that strange birds similar to some kind of waterbird were seen with regularity in the communities on the Ohio side of the river, and also from some other towns near Point Pleasant – and they are still taking place. Some of the original sightings of Mothman refer to it as a 'giant shitepoke' – shitepoke being the colloquial name for any sort of crane or heron – and, of course, the sandhill crane is one of the favorite identities proposed by skeptics. Stan Gordon investigated the sighting of an eerily Mothman-like entity – though one still undeniably avian in nature – at Clendenin, West Virginia in 2004. It is interesting from a Fortean standpoint that the Sproul State Forest, location of many of Pennsylvania's Thunderbird reports, contains a waterway called Clendennin Run. In short, the sightings of weird birds appeared to surface from everywhere *but* Point Pleasant. Could Mothman have been a southern-ranging Thunderbird? Witness descriptions of Mothman do give him attributes similar to the mysterious birds, such as coloration (usually described as gray, white or brown). He also has a propensity for lurking along roadsides (shared with many other "monsters" as well; this also would make logical sense for a large, gliding winged creature and has been noted in some large raptors as they use the air currents stirred up by the passing vehicles to aid them in taking flight), and the size of the Mothman generally tallied with the Pennsylvanian reports. Mothman just had that pesky humanoid form thing going on... but with regard to that, it is interesting to note that many Native American tribes see the Thunderbird as a giant who dons a feathered cloak to become a huge bird.

Another piece of the Mothman-thunderbird puzzle was provided, in part, in 2001. Pastor Robin Swope reported seeing a large bird in a graveyard in Erie that July. He reported that the creature had a wingspan of between 15 and 17 feet. It had a long, thin bill and no apparent neck. The bird was black or dark gray with a darker circular patch on the throat. On his blog, he compares a sketch of what he saw to a blue heron in flight and the resemblance is remarkable, save for the difference in size.

Strangely, a few days before the sighting by two Point Pleasant couples, the Scarberrys and the Mallettes, of the creature that was to become known as Mothman, five gravediggers working in West Virginia were to encounter a brownish man-like form flying in the air over a cemetery near Clendenin. This was on November 12, 1966.

A little over 22 years previously, in July of 1944, Father J.M. Johnston of Hollywood, Maryland saw a winged humanoid form, which again took roost in a cemetery. Why should winged creatures have this sort of propensity for showing up in cemeteries and revealing themselves to men of the cloth?

But we digress.

Pennsylvania's mystery birds are a relatively straightforward phenomenon of sightings and little if any supernatural attributes, but these undeniable parallels with the Mothman and thus, possibly with other winged humanoids seen worldwide, complicate matters immensely. The birds do seem to be of a fairly mutable appearance and though many birds so indeed change coloration as they age and go through molt cycles, it's also pretty apparent that some of these sightings are misidentifications of known species. It seems that herons and similar wading birds are among the most commonly misidentified – and a spur-of-the-moment identification of, for example, a great blue heron (*Ardea herodias*) in flight as something unidentified or prehistoric can be forgiven, as anyone who's seen a flying heron can attest! However, the intriguing amount of consistency in many of the descriptions suggests that there is some population group – even if a localized mutation of something known rather than a separate species – of large predatory bird frequenting the waterways of the northern portion of the state.

These parallels with Mothman notwithstanding, winged humanoids are lacking from amongst our weird menagerie. There is, however, one notable exception: the Granite Run Gargoyle. It seems that in the summer of 2007 (probably June or July), a man had stopped to fuel up around 1:30 AM near Granite Run Road in Lancaster (Lancaster County). Through later discussion with the original investigator, Rick Fisher of Columbia, I discovered that this sighting took place at the Turkey Hill on Manheim Pike (below). After hearing a scratching noise (a

detail often described during sightings of winged humanoids), he noticed a 3-4 foot tall hunched shadow on the roof of a nearby building. The featherless wings were raised above its head. The witness stepped away from his original location, and when he returned to his vehicle he heard the flapping of wings and saw that the creature had flown to the roof of another building. Rick felt that the individual was credible and not deceitful, and the witness made clear to him that it was not any normal animal with which he was familiar.

The next creature in our winged menagerie overlaps into water monster territory. The story comes from Grunderville (Warren County), 1906. This town was only a lumber mill and associated buildings, and the mill was only in existence for six years (1900-1906).

> Persons at Grunderville, three miles below this town, are excited over the appearance in the Allegheny river of a strange water monster in the form of a serpent with wings. Miss Rachel Talbot, daughter of W.A. Talbot, who has a summer villa opposite Grundervllle, was first to see the creature as it came swimming up the middle of the river, the head protruding several feet above the surface. She called to "Hank" Jackson, ferryman for the Warren Lumber company, who ran for his rifle and opened fire. Immediately the reptile reared its head at least 10 feet in the air, Jackson says, and charged for the shore, its eyes, as big as saucers, fixed on him. Jackson steadied himself and, taking careful aim, the bullet hit one of the wings, disabling

it. Jackson says the snake finally flew as high as the ferry cable, which hangs 20 feet above the water, and then vanished.

It was reported that Arthur Savage, of Canal Township (Venango County) captured a flying snake with "a fine coat of feathers and a head like a chicken" about three feet long, which ate grain, in 1900. Ivan T. Sanderson, the famed cryptozoologist and good friend of Bernard Heuvelmans, wrote an article in the Blairstown newspapers detailing an experience he had at his farm near Columbia, New Jersey in 1965. Here, Sanderson encountered what he called the Wooo-Wooo along with two friends, Walter McGraw and Tom Allen. The three men were outside when all nearby wildlife sounds ceased. There was an owl-like hooting sound – which gave the creature its name - advancing along the hills of the Kittatinny Ridge. After a few moments, Sanderson heard a second, similar hooting from across the Delaware in the direction of Bangor, Northampton County and then the sources of the two sounds moved towards the south. Sanderson soon after spoke with no less than four other people who had heard a similar hooting cry west of Bangor, near Saylorsburg, in Monroe County.

Across the river from Sanderson's home and northeast of Bangor is the area known as the Delaware Water Gap (below). Poconos-area researcher Chip Decker (now deceased) also referred to local legends of a similarly-described noisemaker he called the Delaware Water

Gap Hooter, whose haunting cries were heard from among the rocky forests on the Pennsylvania side of the river. It was reported that other animals would be silent during the Hooter's cries, and that people in the vicinity during the sound could feel an odd sort of "heavy" feeling. The Hooter, whatever it is – no-one can correlate the sounds with any known animal – is clearly the same phenomenon as Sanderson's Wooo-Wooo.

Really, the sole reason I am mentioning the Wooo-Wooo/Hooter phenomenon in this chapter was Sanderson's feeling that the sounds were made by a large owl-like bird he saw flying by at the time he heard the hooting.

The last report we shall discuss here takes us to Chester (Delaware County); it seems that some

> "...large bird or bat arose from the Tinicum swamps early today and crashed head-on into the car of Mrs. Ida King, of 121 Central avenue, she reported. The "thing" was so large that it smashed the glass in the car to pieces, and left a mixture of feathers, fur and blood smeared over the car."

# Chapter Seventeen
## Werewolves

The *loup-garou* encountered in France (the name, by the way, is somewhat redundant; *loup* means 'wolf' and *garoul* or *garulf* means 'werewolf'; so 'wolf-werewolf.' Alright, we get the point, he's a werewolf) seems to be very much of the typically European phenotype – heretic goes into the woods and makes a deal with the Devil, puts on a wolfskin or uses some kind of ointment, thereby transforming into a lupine form and becoming a general hellraiser. Just witness the Werewolves of Poligny, or Jean Grenier, or Gilles Garnier (and, weirdly, names like Grenier and Garnier seem to linguistically be derived from a similar source as *garou*)…

However, once the French got to the New World and their werewolf traditions were mixed up with Native American legendry, the *loup-garou* becomes a heady mixture of both. Most French were concentrated in Canada, Michigan, upstate New York and, of course, Louisiana and most of the Mississippi Valley. Here the *loup-garou* is not only capable of turning into a wolf. Calves, oxen, pigs, cats, owls, horses, and apes are all alternate animal forms for these shapeshifters. Sometimes, the New World *loup-garoux* are only doomed to change shape for a limited time (usually 101 days), unlike the European werewolves, which were usually cursed until their deaths, which also usually came at the hands of pissed-off townspeople.

And I suppose that I really shouldn't be surprised that some stories which seem to be of *loup-garoux* are found in Pennsylvania. There's any number of ways we could have inherited the French wolfmen. Some of the Pennsylvania Dutch, after all, came here from New York state and it's not at all unlikely they brought some of the stories with them. Detroit, Michigan, which has been noted to have many werewolf stories, isn't all that far from, say, Erie (and my own family is present in Michigan, where the name Gable has unfortunately been associated with the hoaxed Michigan Dogman video). During the French and Indian War, many French forts existed in western Pennsylvania. Pittsburgh, itself, was formerly Fort Duquesne.

In fact, I suppose, there may be some justification for the idea that most of the werewolf tales in northern Pennsylvania are more or less of French origin. It has been noted that many

## The Mystery Animals of Pennsylvania

French-Canadians were migrant workers in Pennsylvania during the logging boom in so many of the northern counties. The term "spook wolf" seems to be generally used as interchangeable with werewolf and seems to have been used to describe the wolf form of these supernatural creatures, although a number of spook wolves also seem to be more-or-less regular wolves given supernatural attributes.

Probably not too surprisingly, since Henry W. Shoemaker was a resident of Clinton County for a long while, many stories from that mountainous region of the state are known.

### Traditional werewolves

Peter Pentz of McElhattan in Clinton County, an informant in many of Shoemaker's tales of the northern mountains, told of a werewolf which frightened his great aunt, Mary Depo. A huge black dog approached her, rising up onto two legs as it did so. Mary ran all the way home, the bipedal dog chasing her. Her husband snatched up two *pewter* bullets coated in sacred wax (shades of the way the Gifford's Run werewolf discussed later was slain) and shot the apparent werewolf with them. The wolf sank to the ground, transforming into a neighbor almost before the man's eyes. The neighbor thanked him for freeing him from the lycanthropic curse and died.

A French engraving known as Les Lupins; Mary Depo reputedly encountered a creature much like these sometime in the 1800s.

The gratefulness of Mary's wolf, however, is very much in keeping with certain Italian werewolf tales; and the curing of lycanthropy by causing the afflicted person to bleed, which Shoemaker likewise mentioned, is common both to the Italian stories and the stories of the *loupgarou* in Canada. All of which is a convenient bridge to some other tales, since Jacob S. Quigley (the name is given as Quiggle elsewhere), a great help to Shoemaker in gathering the Clinton County werewolf traditions, was of Italian heritage.

Many of these surfaced from Wayne Township, a mountain district south of McElhattan. George Quigley, Jacob's father, shot a large brown wolf in the foreleg one night on his farm using a silver bullet. It uttered a woman's scream, jumped over a stone wall, and vanished into the bushes. A short time later, an old woman of the neighborhood walked into a neighbor's cabin cradling a broken arm, which she claimed she had broken while shooting at a fox.

His mother likewise mounted a watch at the farm of a neighbor, who reported that she had found her horses tired and weary every morning. It was eventually determined they had been taken to Stone Mountain, *Falsbarich*, where witches were rumored to hold their Sabbats. Mrs. Quigley found that black wolves were entering the stables and riding the horse towards the mountain. So she and a few other neighbors covered the saddles with the thorns they had found tangled in the horses' manes, and the black wolves uttered fearful cries when they sat on them. The next day, an old woman in the neighborhood claimed that she had landed in a bed of nettles when she was thrown off her horse.

I'm not certain whether the first of the Quigley tales is the same as that recorded to have happened to a George Wilson, as the details are nearly identical. When mentioning that case in *Wolf Days in Pennsylvania*, Shoemaker mentions that Wilson also killed a three-legged wolf on his farm with a silver bullet. (At least, apparently, Wilson was real: a 'G. Wilson' owned a farm that is now in the inaccessible forests off the modern Pine-Loganton Road, as of 1862.) Both Grimm and Baring-Gould mention that at a certain hour of night, werewolves were thought to assume a three-legged form.

Shoemaker heard some werewolf tales from an old man in Frenchtown (Clearfield County). The old man, with the awesomely French name of Bilotte, told of the killing of a werewolf near Gifford's Run sometime back around 1845. This area, about four miles north of Frenchville, is also known, ominously, as Devil's Elbow. Much has been made of how many places christened with the name *Devil's* ---, in the United States at least, are actually based on a mistranslation of the Native American term *manitou*, spirit. Gifford's Run is also the site of rapids, and thus is prime territory for representations of the water panther. Mr. Bilotte's werewolf was killed with a silver bullet wrapped in sacred wax, and after death changed back into a man who was known to be an *émigré* from the Provence.

A well-known tale comes from Northumberland County. It is the tale of May Paul, whose family owned a farm along the Schwaben Creek near what is today the town of Rebuck. An old hermit who lived nearby would sit and watch the 12-year old May. Her family didn't really approve of the old man and his habits, but since he apparently kept the wolves away, they didn't bother too much.

One night, a traveler saw a big wolf crossing the road. He shot the wolf, which crawled off into the woods. However, the trail of blood led not to a dead wolf, but to the old cabin in which lived the old hermit. The old man was buried just outside his cabin, and that area of Line Mountain is still known as *Wolfmansgrob* (werewolf's grave). Despite the wolves that ravaged nearby flocks, May never lost a single sheep to the predators. She claimed the ghost of the lycanthropic pedophile still came to sit with her.

## Meneurs des loups and other wolves in human form

If French-style werewolfery was recorded from Pennsylvania, I suppose we shouldn't be surprised that an associated tradition also is found here. This is the figure known in French as *meneur des loups* (leader of wolves), a wizard who could "speak the language of wolves" and command the werewolves in their depredations.

Indeed, there are a number of men in legend who claim to have been able to speak to wolves. A few examples of this unorthodox hunting style are given by Shoemaker in *Wolf Days in Pennsylvania*:

> "One of the most famous wolf hunters in Pennsylvania was "King" (Henry) Heizmann, "The Bear Trapper," who died near Boyersville, now called Mazeppa, Union County, in 1895. Every fall this eccentric man... [would] go to the White Deer Mountains where he trapped until spring. He captured many wolves, luring them out of the forests by imitating their cries. Edwin Grimes killed wolves the same way in McKean County... "Bill" Long, also called "The King Hunter," as a small boy surprised his father by "calling" wolves out of the forest to be shot. He learned the trick, he said, from friendly Indians, who frequented the elder Long's still-house."

Elsewhere in the book, it is stated that Robert Quiggle of Clinton County "told of his brother, William Quiggle, who used to climb a certain big shellbark back of their home and bark like a wolf, drawing many brown wolves off the mountains close enough to be shot."

The pastoral region of Warrington Township in northern York County has spawned a few supernatural tales. One 1875 tale, recounted by author Thomas White, told of a specter haunting a Mrs. Nesbit that was laid to rest by an act of sympathetic magic prescribed by a powwower – this act seemed to be not out of place in a werewolf tale.

Perhaps, then, it is not surprising that quite possibly the only werewolf legend in York County comes from that place.

The Ross family took in a mysterious girl who appeared at the Warrington Meeting House one day. The girl eventually began to make wolf-like noises in her sleep, and was soon suspected by the Rosses to be a *garol* or werewolf. She vanished one night, and though the ground was muddy, the mystery girl could not be tracked. No human tracks could be found, although wolf tracks were. And the Rosses shook their heads, for she was a werewolf, and now she was with *her* brethren.

## Other shapeshifters

It is noted that the Oley Valley of Berks County was settled by Frenchmen (Huguenots, French Protestants, to be precise; just the kind of folk that would likely be labeled heretics and prime for werewolf transformation), similarly to Clearfield County; are there were likewise any werewolf traditions there? Well, there are, but they're not as much a werewolf as another sort of shapeshifting *loup-garou* – and it is related, albeit tenuously, to the infamous Rehmeyer's Hollow affair. The murderer's ancestor Jacob Blymire, a "seventh son of a seventh son," was born on the same day that the Oley Valley's famous witch, Mountain Mary, died (in November, 1819). He had a "strange fondness for owls," according to John's father. And as noted in David Kriebel's extremely thorough *Powwowing in Pennsylvania*, John believed that the curse under which he was suffering was placed on him by, among others he blamed at various points, the ghost of his ancestor Jacob. John heard a barn owl hoot seven times at midnight one night and that, of course, meant... the curse-layer had to be Jacob? Yeah, the logic doesn't quite follow. But as John Blymire's – to put it politely – *lack of intelligence* was noted at the time of the murder, I guess logic shouldn't really be expected.

During her life in the Oley Hills, an owl consistently visited Mary's farm, drinking the milk as soon as she got it out of the cow. One evening, she burnt the feet of the owl to keep it from returning to plague her farm – and the next day, she heard of an old woman on a neighboring farm whose feet were burnt.

Another Berks County tale akin to those of the spook wolves is known to the curators of the raptor sanctuary at Hawk Mountain. Called the Great White Ghost Bird, it frequents the area of the old Schambacher tavern. A psychic who visited the mountain seemed to indicate that the bird was some sort of spirit of vengeance, being responsible for the death of a murderer in the 1880s. Seth Benz, an inhabitant of the tavern (now a private residence) described the bird thusly:

> "It takes on different shapes. Sometimes like a pigeon, sometimes a goshawk, a falcon. It's a mysterious thing... I was up on the Hemlock Heights one day and I saw what I thought was the same bird. This autumn, one of the interns here spent some time in the house. He saw the ghost bird, too."

Interestingly, an "albino ape-like creature" is noted as having frequented the area of Bethel Township, in the foothills of the Blue Mountains just west of Hawk Mountain. Could this actually be yet another form of the white, ghostly bird? (There are also tales of other wildmen further east, in the hills above Shartlesville.)

Lehigh County is the point of origin of two tales of shapeshifters of another sort. The northern sector of the county was terrorized by a vicious black pig that would attack travelers on the road. One traveler tossed a log in the direction of the pig, smacking it on the leg and breaking it. It was later learned that an old woman who lived in the vicinity had broken her leg. The monstrous swine was never encountered again.

Meanwhile, Trexlertown in the southern portion of the county was supposedly the abode of an old witch who was offended in some manner by the wife and three daughters of a farmer named Weiler. A curse was placed on the Weiler household by the witch. For three months' time, anytime a visitor came to the house, the daughters would be transformed into serpents, crawl around the room for several minutes, and eventually change back into girls.

A house in Schuylkill Haven, Schuykill County, was troubled by a ghost – or so it seemed – in January, 1917. The standard things that are usually caused by some kind of supernatural malevolence took place – chickens refusing to lay eggs and milk would turn sour after only a short time. Weird noises were heard and the shutters on the windows refused to stay closed or open, however the homeowners left them. An article on the event in *The Call* (January 17, 1917) notes:

> "During the week the cause was discovered and undoubtedly removed. A pigeon was discovered making its home in the garret of the house and with a well pointed shot from a gun that had been loaded with only thirteen small shot, the pigeon was killed, one of the shot having penetrated its neck. The following day a certain resident was noticed wearing a bandage about their neck. The supposed hex is not truly a resident of Spring Garden, but a person who lives on the outskirts of the borough and who daily makes trips to all sections of the town."

Schuylkill Haven, by the way, is only a three miles south of the Tumbling Run Valley, location of the Hex Cat saga (1911), so clearly hexes and shapeshifting were very much the fashion *du jour* in Schuylkill County.

Christine Schaefer recounted the tale of a shapechanger of another sort that was encountered in her apartment on East Gibson Street, in Scranton (Lackawanna County). The manifestations of this creature began when her father tore down a wall in the apartment's bathroom. Shortly thereafter, the phenomenon began, innocuously enough, as an open door. Her parents, not having opened it, would shut the door, only to have it open again. Then, in the autumn of 1950, Christine would be witness to a strange apparition: an emerald-green fox, grinning eerily. An exorcism was called for: but, as is often the case, it had little or no effect. Christine still saw the Cheshire Fox, and the more it appeared, the more it grew. Some of the British black dogs can change their size like this, as did other Pennsylvania zooforms like the "hex cat."

Almost a half-century later, Christine's mother would recall that she also saw the green fox: but, in her recollection, it was a fox-headed child. Did it appear differently to the two women, or did Christine merely misremember it? We'll probably never know, as the apartment building has been torn down. (It should be noted, though, that the lot where the apartment building once stood is only a matter of blocks from where the first sighting of the Woman in Black was reported back in 1886.)

The mention of a fox begs comparison with a prominent legend in southeastern Pennsylvania, that of a werebeast called Red Dog Fox. Tales of this creature come from the Brandywine Valley, a region encompassing portions of Chester and Delaware Counties, as well as northern Delaware. A young red-haired man named Gil Trudeau was a guide for the Americans during the Revolutionary War. Cries like those of a fox were heard around the soldiers' camps, dead animals were found near them, and occasionally blood stained Trudeau's clothing. Some men saw a huge, red fox following them. Finally, the beast was shot and – as you should be able to guess by now – turned into the corpse of young Trudeau. The ghostly form of the Trudeau-fox still haunts the valley. Eastern Chester County fell both within the aforementioned valley and the Welsh Tract, that region of the state settled by Welshmen, and possibly preserving some of their traditions, discussed in an earlier chapter. Marie Trevelyan records that in some parts of Wales, "sandy-haired people and those with dull red hair [were believed to be] descended from foxes." Did this contribute to the tales of Red Dog Fox?

Just as I was finishing writing this book, word came from England that a gigantic fox, nearly four feet in length, was killed near Maidstone, in Kent. So foxes as large as the one in this story definitely *do* exist.

Finally, from Wolf Camp Run in Bedford County comes the tale of the 'wolf-wife.' Similar to many tales of mermaids, selkies, and other folkloric creatures, this was the tale of woodsman Harman McNeel, who tracked a black wolf to an old mill-house. Here he found a beautiful woman, whom he later married, but who was actually the wolf in human guise, keeping her form only as long as her wolf's skin was kept hidden from her.

## Criminals of the lycanthropic kind

Finally, it should be noted that lycanthropy seems to have some odd sort of fascination for Pennsylvania's criminal element. The first real criminal case in which lycanthropy was involved – supposedly – came to light on March 29, 1940, when forty-year old Michael Menichella of Philadelphia's Tacony neighborhood confessed to murdering his baby daughter Beatrice and tossing her body in a coal bin in his cellar. Previously, he had insisted that he was not the culprit in the murder, but "witches and a big black dog" were. But the fact was, after he confessed to the murder, the story only got weirder. He still maintained that "witches made him do it." Menichella continued, "A big black dog came in through the window last night. To protect myself, I turned into a dog. Anything from there on must have been done by the dog."

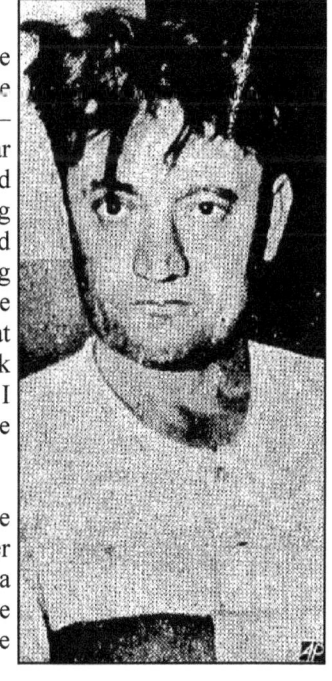

There was another scandal going on in Philadelphia at much the same time involving witchcraft. Morris Bolber was a former New York grocer who was known as "Louie the Rabbi", a "veteran witch doctor and compounder of charms," and Rose Carina a dour, black-clad woman with a "strange power over the superstitious" who became known as the "Kiss of Death Woman" (this owing to the fact that three of Mrs. Carina's husbands had died under suspicious circumstances). At some point the two came to Philadelphia and set up their own criminal organization. They were to make national news in the late 1930s after Bolber's arrest in May, 1939 and Carina's arrest in New York nearly a month later. It came out in the various trials that "a lot of people called Morris Bolber 'God'" – nothing like humility, I guess. He himself testified against another sorcerous member of the so-called Poison Ring, Paul Petrillo, but all was for naught; Bolber was sent to Philadelphia's infamous Eastern State Penitentiary in 1942. Could these have been the witches and hexcrafters Michael Menichella went on about? The Menichella case could likewise have been simply an ill-thought out attempt by a criminal to capitalize on the 'hex hysteria' which had already led to four murderers – John Blymire, John Curry, and Wilbert Hess in 1928 and then Albert Yashinsky in 1934 – to be imprisoned, rather than executed. It must have worked, because Menichella didn't die until 1968.

Menichella's is the most blatant case of supposed lycanthropy among Pennsylvania ne'er-do-wells – the rest appear to have been mundane (though no less severe) offenses with a 'werewolf' element linked tenuously.

The first one is the least severe, a mere nuisance, really: Craig Brashear, a young man from Enola (Cumberland County) was charged on a summary offense after he admitted that he was the East Pennsboro Monster. This was a creature seen in 1985 in East Pennsboro Township, the region around Enola. The young man ran through fields and roads near Enola, wearing a "monkey suit." However, a photograph of Brashear's outfit clearly shows that the mask was much more 'werewolf' than 'Bigfoot.' (The first sighting of the East Pennsboro Monster, though, was a genuine report and definitively *not* the work of Brashear.)

Another "werewolf criminal" was Vaughn Crews, a young man from Donora (Washington County) who in 2008 broke into a woman's house and raped her, wearing a werewolf mask as he did so. The wearing of a mask is nothing new – criminals often conceal their identities – though the choice of a werewolf mask is one I haven't heard of before. Crews, by the way, has received 17-40 years in state prison.

Another 2008 case – which occurred only two weeks after Crews' offense – was a case in which Kristian Carl, a troubled young man from Pottsville (Schuylkill County) was charged with statutory sexual assault after he managed to convince a 15-year old girl that he was something straight out of the *Underworld* movies. (Not Kate Beckinsale.)

> "He convinced himself he was a hybrid – a combination werewolf and vampire," Pottsville police Sgt. James Joos said. "He had convinced the girl he was, too."

To prove to police he was indeed a genuine vampire/werewolf, Carl "showed me his canine teeth," Joos said. "I let him know that all mammals, including humans, have canine teeth." Joos said Carl also told police he had a "guardian dragon that protected him from evildoers."

# Chapter Eighteen
### Black Dogs and Phantom Hounds

A common thread of folklore in many European countries are stories of those black dogs known by many names in different cultures and particularly known from the British Isles. The dogs are often associated with running water and equally often with places of death – be they places of execution or cemeteries. Some localities have a tradition that the dog will follow wanderers on lonely, moonlit roads and that they herald death to the one who spins round to confront them. In fact, the British stories of black dogs are too complex to deal with in a more than cursory manner here.

We'll start with another story from Berks County's Oley region – Lobachsville, to be specific – tells of lost treasure and a guardian demon.

The Keim family of that place were rumored to have had a treasure worth more than $100,000 secreted somewhere on their farm. A neighboring farmer told the tale of a black dog with yellow eyes that chased horses and was seen wandering the roads. Once, it chased a horse, vanished, and reappeared to bite the horse's throat with its ghostly jaws. The horse died the next day. The farmer was later told that the spot where the dog rose from the earth was the very spot where the Keim family treasure was hidden.

But closer to Oley proper, the tale is of a similar being – but one apparently more in line with the bipedal man-wolves Linda Godfrey has recently done much research on. This variant is noted to be a similar black dog, one with "yellow Satanic eyes" that generally appeared to be about the size of a Labrador. However, the dog could rear onto its hind legs to give voice to a terrible howl and when it would do so, it grew to nearly ten feet tall.

The football field of Reading Central Catholic High School, near the Oley Hills town of Jacksonwald, was once the site of an old farmhouse where folk claimed to have seen the form of a "fiery man running up a tree" and others saw fiery dogs. No word on whether they meant these were glowing forms or whether they were actual hellhounds. Interestingly, some Welsh traditions of the *cŵn annwn* or *cŵn wybr*, the hounds of the Wild Hunt, have them as "flame-coloured," and another describes a similarly bipedal hound with eyes "like moons" which haunted the area of Wilton, in Glamorgan. We have already established that at least the tradition of the "corpse-bird" survives in Montgomery County, a remnant of the Welsh settlement of the area; who's to say others didn't likewise endure?

Not leaving the Oley area, the estate of Johannes Lesher, the ironmaster of Oley Forge (now an art gallery) is haunted by the form of a black dog – which often accompanies the form of a tall man, usually said to be old Lesher himself. The Fortean – and particularly spectral – lore associated with iron furnaces is well-established.

A forest in Montgomery County was known to be haunted. Particularly active was a bridge along the Newtown Road, crossing the Pennypack Creek, in present-day Moreland Township. In October of 1790, there were several encounters with a hellhound in the vicinity of the bridge. A man by the name of George Layton was walking along the road when he saw a bright flash.

> [He] observed a flash of lightning and on looking around beheld what he imagined to be a huge black dog with eyes of fire, dragging a long heavy chain whose clanking he distinctly heard. Extraordinary to say his eyes appeared to be the size of his mother's pewter plates.

One man "saw some monster following him that by the sound appeared to be trailing some huge chain on the ground," and the account of a young man from Crooked Billet (now known as Hatboro) was given:

> "As he entered the woods, on account of the density of the foliage he soon found it exceedingly dark at this dread hour, and was ad-

vancing at his usual rate, when on a sudden, near the roadside on his right, a loud snort was heard. He looked in that direction, but did not see anything. Shortly there was a louder snort. *But to think!* distinctly and approaching came the loud, clear clanking of chains; and again, nearer, a still more demoniacal snort, with louder snorting, that seemed to vibrate throughout the woods... He started at full speed, and to his terror was as rapidly followed by the distinct snorting and continuous clanking, till the very ground appeared to tremble under the sound."

Another man, named Peter Pennel, said that he witnessed a tall man, dressed in black, near the haunted bridge. The ability to transform from a canine to humanoid form is often alluded to in tales of European black dogs, particularly in the Low Countries; and some traditions associated with the East Anglian spectral hound Black Shuck have it making appearances as a dog-headed monk – specifically, those near Clopton Hall in eastern Suffolk.

The mention of huge eyes and the sound (or sometimes visual) of a dragging chain often appears in encounters with British black dogs, particularly a pair of spectral hounds appearing near Leeds in northern England, Barguest and Padfoot (which names may or may not be interchangeable). Barguest, at least, is also often described as headless – a feature of another Montgomery County legend.

About a mile southeast of Schwenksville, the West Skippack Pike spans the East Branch of the Perkiomen, near where it joins the Perkiomen Creek proper. On the southeastern side of the East Branch (sometimes simply called Branch Creek) there once was a large farmhouse which was used as a tavern by the "grim and sinister" Jacob Ellinger, a blacksmith, and his family. The family was said to have consisted of Jacob, his wife, and five children (four sons and a daughter). It was said that Mrs. Ellinger was a *femme fatale* of the truest sort  she "would persuade the peddlers to her embraces and then at a certain signal [Jacob] came and robbed them." The Ellingers also stole chickens – and robbery was only the least of their wrongdoing.

There were persistent tales of people who stopped at the Ellinger tavern and never returned. They were also horse thieves of the worst order, killing travelers and stealing their horses. Once, they murdered a traveler and threw his body in a well, sealing it up over the corpse and digging a new one.

William K. Ellinger, who was later to become a prominent member of the Mennonite Church in Montgomery County, was actually one of the four sons of the ill-famed clan.

But stories of ghostly happenings continued for years after the Ellinger place fell into disrepair, eventually becoming just a ruin. It is often said that the grounds are roamed by a headless man mounted on a white horse, said to be the remnant of Reuben Boorse (also sometimes called Jacob Bucher), the man sealed in the well by the robber clan. Sometimes, this is foregone and the ghost shortened to merely a headless white horse. Lights were often seen in the ruin, and one man saw a "rolling ball of fire" come down the hillside towards the road; he dove into the waters of the Perkiomen.

It was likely in the 1870s or thereabouts that a man and wife were driving their wagon past the ruined Ellinger tavern. The horse came to a halt. Then, as Governor Samuel Pennypacker (whose home was located a short way to the north) recorded:

> "A huge white dog came out, reared upon its hind legs, and with wide-open jaws, thrust his head forward into the wagon as if to seize them. The man, frightened at the unearthly spectacle, whipped up his horse. They then looked back, saw the specter still standing upreared when it suddenly faded into mist and, without movement, disappeared."

The apparent aggression is similar to the tale of the Black Dog of Lobachsville. Similarly, the mention of its bipedal stance should be noted in conjunction with this, since Linda Godfrey has mentioned that the "man-wolves" of the Upper Midwest often display a similar sort of pseudo-aggression, apparent attacks that never *quite* connect.

Around 1888, Charles H. Tube and his wife were returning home from Schwenksville when an animal appearing to be a headless calf emerged from the Ellinger ruins and approached their wagon. It followed them for a distance. In 1890, a man named Frank Ziegler was riding near the ruins when the white dog reappeared – a "huge, white dog" stepped onto the road from the ruins and simply "melt[ed] into shadow." No mention of any bipedal stance this time, though it *is* noted that Ziegler's horse froze as did the couple's horse 20 years before. Another man, by the name of Henry Wireman, was pursued by a "phantom with eyes as large as plates" near the ruined tavern around 1897; Frank Wireman also saw a man driving a cow and a calf, and all three were skeletal.

While there isn't much to directly connect the beasts reported to the black dog legends of Britain, it is nonetheless interesting to note the "eyes as large as plates" of the Wireman phantom in relation to those on the Moreland Township creature. The murder of a peddler, and its connection with the appearance of a black dog, is likewise mentioned in the tale of the Druggen Hill Boggle in Cumbria. A headless appearance is often noted in the phantom realm, apparently having little to do with whether the poor unfortunate was actually beheaded or not.

Interesting indeed, that the sole Montgomery County report received by the Pennsylvania Bigfoot Society is of shrieking vocalizations heard throughout the 1990s on a farm near Schwenksville. Is this apparent Bigfoot yet another shape taken by the Ellinger phantoms?

Neighboring Schuylkill County is also home to at least one black dog, which haunted Seven Stars Road. In the 1950s, a man named Theodore Ebert told George Korson about an encounter that had taken place years earlier: "a big black dog appeared from nowhere and came between me and one of my pals. And I went to pet the dog, but it disappeared from right under me. Just like the snap of a finger it disappeared."

Ebert was just a boy when this black dog encounter took place; a young boy near Columbia in

## The Mystery Animals of Pennsylvania

Lancaster County had a nearly identical encounter one evening in the autumn of 1944. Just opposite the gate to the property of poet Lloyd Mifflin is a lane leading to a house. It was along this lane that the boy saw a large dog, rather like a wolf or German Shepherd, walking and emerging into Norwood Road, sitting down at his feet. He reached down to pet the dog, but his hand went right through the image. The dog stood up, passed right through the wheel of the boy's bicycle, through the gate of Norwood Mansion, and vanished.

Strangely enough, the next year the same boy saw the form of a white-clad woman standing in Norwood Road at almost the same spot as he had encountered the dog, walking through the same gate and vanishing. Was it connected somehow?

### The black wolves of the Seven Mountains

Centre County, in the northern part of the state – and, to a lesser extent, the neighboring Union County – have melded the black dogs of folklore with the body of lore concerning the native wolves to create a monstrous hybrid creature. This hybridized nature of the black wolves is an ironic twist, since it is extremely likely that black wolves, in reality, are descended from dog/wolf hybrids.

Doubly ironic, given the last case we discussed, that the discussion of black wolves will be started with discussion of the case of William Etlinger. *Triply* ironic, that a *Pittsburgh Press* article calls Etlinger a "burly blacksmith," given the occupation of the shunned Montgomery County clan. In any case, the fellow has undergone a number of name changes over the years, the first recounting of the ghostly legend associated with him appearing in the works of Henry W. Shoemaker in 1909's *Pennsylvania Mountain Stories*. The story was deemed to have happened too recently (occurring only 13 years before), and by Shoemaker's own admission many of the names associated with the story were changed, with the name of the antagonist being given as 'Silas Werninger.' By the time of that same author's 1952 essay "The Werwolf in Pennsylvania," his name was changed to 'Cyrus Etlinger,' and the scene of the events was given as Woodward (Centre County). How could the wrong first name and the correct last name have been associated with the man? An 1862 map of landowners in Centre County lists a farm north of Woodward (on the aptly named Cemetery Road) as belonging to a 'C. Etlinger.' Was Cyrus actually the father of the antagonist of the next tale?

Armed with the knowledge of a 'scene of the crime,' so to speak, I uncovered the story. William Etlinger was, by the accounts of nearly everyone in Woodward, "a man whom it was not best to provoke." He had struck his wife's father, Benjamin Benner, with a log and nearly killed him. He was arrested, but freeing himself on bail failed to appear in court on the appointed date. A Constable Barner, along with C.G. Motz and John Hosterman, went to the farmhouse on March 5, 1896 to take Etlinger into custody. There, he killed Barner and seriously wounded Motz. He then barricaded himself, his wife, and two children in the house. After a lengthy siege, during which Etlinger shot several townsfolk and very nearly the sheriff of Centre County, the determination was made to set fire to the house. Soon, his wife and children exited the structure and as the house burned, William Etlinger shot himself.

The scene after the crime was an odd one:

> "The brother and the widow of the outlaw refused to have anything to do with the body and the overseers of the poor took it in charge, placed it in a hastily constructed coffin and buried it in the mountain. It was hauled in a farm wagon half a mile east of the village, and laid in a lonely grave in the Narrows."

On March 9 came word that the body of William Etlinger had been stolen – and to this point, the story tallies well with Shoemaker's version, even though the names were changed. His version had it that after his death and burial in the mountains, the outlaw's grave was frequented by a gigantic black wolf. After consultation with a local witch, it was determined that the black wolf was the Devil himself, come to claim the soul of the renegade, and that the plot could be defeated by exhuming the corpse and reburying it in the Lutheran Cemetery in town. This was done, and the black wolf gone.

But the next day came a retraction – no, the body was not gone, it was still in its mountain grave. A few days after the events at the Etlinger house, a young girl in town died under mysterious circumstances. So far as I am aware, the cause of death was never determined, or at least not made public. A dreadful possibility presents itself – according to the testimony of his wife, Etlinger "had also secured a large amount of poison with which to poison the waters in Woodward to get revenge on his neighbors" – could the young girl have been his daughter, poisoned by her father before she left the house?
**UPDATE:** In May, 2011 it came to my attention that, in addition to the names, both Shoemaker's original, and subsequent, accounts of the incident changed the details of the associated ghost story as well. I've since learned that the tale was *not* that of a ghostly wolf, but of the eerie image of Etlinger's cabin, alight with flame in the forest near Woodward even years after it had burnt to the ground; but similarly to the details Shoemaker recounts, the body of the suicide has been moved from the woods to the town cemetery.

The story of the wolf come to claim the soul, incidentally, is quite similar to an earlier tale related by Governor Stone in his autobiography – there was a similarly ill-reputed man in Wellsboro (Tioga County) named Richard Duryea. If rumors were to be believed, Richard was an old pirate and if even wilder rumors were to be believed, a Satanist who picked fistfights with preachers. One day, he fell ill. John Ainsley and Andrew Kriner sat with Duryea, and they saw a huge black wolf with "sharp" eyes run into Duryea's room and then back out again – and when the two men checked on Duryea afterwards, he was dead.

But back to black wolves of Centre County – they seem for the most part (and quite oddly, for wolves) to be not so much aggressive as unnerving. Several instances are on record of the wolves merely following someone, though always at a distance. The grandfather of J.R. Ramsey was followed by a group of the black wolves as a boy, and began throwing them meat. As they persisted in approaching, he struck a stone against the blade of his scythe, whereupon the wolves ran away. The grandfather said, "The plague take you all! If I had known you liked that sound so well, you should have had it before dinner!" Shoemaker noted that this tale was an almost word-for-word mirror image of an old Scottish legend.

Earl Motz and Charles Hosterman, both of Woodward, recalled a hunt which took place in the 1850s at the Hosterman farm north of town. Here are a number of parallels with the Etlinger story – Woodward, the last names of the two deputies taken to arrest him, and the Hosterman farm was near where the 'C. Etlinger' farm was located. Was this tale an invention of Shoemaker's (a habit he was known to engage in), or is it merely evidence that large black wolves inhabited the area near the Etlinger farm?

> "The wolves, thirteen in number, first appeared in September at the farms in this section. They killed six sheep and wounded a number more the first night. The farmers heard them howl as they went into the woods about daybreak the next morning.
>
> The farmers who lived in the neighborhood, the Hostermans, the Grays, Vonadas and Hinksons, hunted and set traps for the wolves, but were unable to kill any of them. The wolves left in a few weeks and did not return until December of the same year.
>
> About Christmas the pack again came into the valley. They killed a heifer, which had been left in the fields, and devoured a portion of it. Several inches of snow covered the ground at this time, and the farmers tracked the wolves to Hosterman's Gap, and from there east into Pine Creek Hollow. The wolves appeared every night, and all efforts to trap them were in vain. One night two dogs which belonged to the Hostermans followed the wolves and drove them into the woods; there the wolves turned on the dogs and killed one of them and ate about half of its carcass.
>
> The farmers placed the remains of the dog in a tree and the carcass of the heifer on the ground and set traps around the carcass of the heifer. In the morning the heifer was untouched, but the wolves had the snow tramped down under the tree in which the dog was placed in their efforts to get the dog.
>
> At last a trap was set at a place on a small stream where the wolves always crossed when leaving the valley. The next morning the trap held a mammoth black wolf.
>
> After this one was killed the farmers set out poison and the wolves left immediately and never returned afterward. They were seen later in the winter in Brush Valley.

Although many wolves were seen and tracked since that time, that was the last one killed at the east end of Penn's Valley."

Brush Valley, containing the town of Millheim, is just over a mountainous ridge from Clinton County.

## Jim Jacobs and the herald of death

Folklorist Charles M. Skinner recorded the story of the wolf he called the "White Wolf of Venango" – though it seems that the tale happened not in that county, but in Warren County, further to the north.

"On Cornplanter's reserve in Venango County, Pennsylvania, lived an Indian family named Jacobs; big, athletic fellows, full of hard sense and afraid of but one thing: the white wolf. For to see the wolf was "bad medicine"; to chase it, death.

There was never a doubt as to its being a real wolf; it had eaten too many hens and sheep and killed too many dogs to leave room for any question on that point. Yet traps would not catch him; dogs in packs could not bring him to bay; bullets either missed him or glanced from him. A young member of the Jacobs family engaged to guide a party of hunters through this region, and all went well until they had reached the head of the Clarion.

On breaking camp at this spot Jim Jacobs (left) took no part in the preparations.

He smoked a silent pipe and said that the others must go on by themselves; for he had seen the white wolf, and that meant bad luck. They joked and gibed him without moving him in the least. He finished his pipe, told them by what trails they could reach McCarty's trading station, bade them adieu, struck into the forest labyrinth, and went home. He was killed in an accident soon after.

The hunters, scoffing at Jacobs's superstitions, kept on. They got the help of a trapper, who kept a number of dogs, and decided to leave the deer to their liberty for a time and hunt down this hoodoo.

St. Matthew's Lutheran Evangelical Cemetery, formerly the Brungard Cemetery, where the famed white wolf was encountered.

After much luring and watching they came upon the fellow's tracks and on a quantity of pheasant feathers, for he had left his lunch in a hurry, and presently, near Baker's Rocks, they saw him: white as a polar bear, three feet high at the shoulder, bristling and snarling. The eyes of this beast seemed to shoot red fire. Four rifle-shots rang out, and the wolf was gone, with the dogs in hot pursuit. In an hour he was overtaken again, and again the guns were emptied. The animal leaped over a cliff, sixty feet, into a stream, almost at the moment when the shots were fired. No blood was visible, no splash was heard. The dogs found no scent. It was the last time that the white wolf was seen, but in a few months every member of the hunting party was dead."

The Cornplanter Reservation was a 1,500 acre tract of land in northeastern Warren County granted to Chief Cornplanter, chief of the Seneca tribe; when Lake Kinzua (the Allegheny Reservoir) was created in 1965, the reservation was almost completely flooded. Baker's Rocks are also in Warren County, a short distance northwest of the town of Garland.

Jim Jacobs was another of the larger-than-life hunters that were so common in northern Pennsylvania in the old days. He was born Samuel Jimmerson Jacobson near what is now Carter Camp (Potter County) in about 1848, and was the son of a white man and an Indian woman. He claimed to have killed a huge wolf in a snowstorm when he was only six, killed the last elk in Pennsylvania (the elk currently found in the state are actually descendants of ones reintroduced by the Game Commission and not strictly speaking native to the state) and other hunting feats. He was apparently struck, and killed, by a train near Bradford (McKean County) in 1880, but numerous individuals claimed to have seen him on Indian reservations for many years afterward.

Henry W. Shoemaker states that a pair of Senecas named Faddy killed several wolves on the Cornplanter Tract in 1865, and that wolves were quite common on the reservation until 1870. Furthermore, he states that "Jim Jacobs… was conspicuous among the redmen who hunted the last wolves in Warren and McKean Counties." An account is given of a record-sized wolf killed near Kinzua by Edwin Grimes and Ben Main, in 1860.

> "The hide sold to the veteran wolfer LeRoy Lyman who pronounced it the biggest he had seen in his long experience in the forests of Northern Pennsylvania. John C. French, in commenting on the size of the Kinzua Valley wolf, says Edwin Grimes, Sam Grimes, and Ben Main agreed that the grand daddy wolf was at least two inches higher at the shoulder than average wolves and one inch taller than the largest they had ever seen. Continuing Mr. French said: "The wolf was not measured, but it must have been seven feet long without the tail. Ben Main, who was 5 feet 8 inches tall, could not swing the carcass free from the ground by taking an ear of the wolf in each hand and lifting the head at arms' length about his own; but Edwin Grimes, who stood 6 feet 2 inches, could just do so."

## Spook wolves

A tale was also recorded from Jim Depo, who lived in Union County; he said that there was another monstrous spook wolf lairing on Buffalo Mountain (northwest of Tannertown). No bullet could harm this wolf, until a silver bullet which had previously killed a French werewolf was fired at it. No mention if this wolf was ever killed.

The black-clad and long-bearded form of Ben Burkett, well-known as the *Hechsenkeng* (king of the witches), was often seen walking the roads in the middle of the state. He claimed that he had killed a spook wolf in the aptly-named Wolf Hollow, below Ritchie's Knob (Blair County). The monstrous white wolf would kill large numbers of sheep, but drink only their blood. Burkett said he first drew a magic circle in a farmyard it frequented, and then shot it dead with a silver bullet coated in his own saliva. The residents of East Freedom rejoiced when he brought to them the skin of the wolf.

It should be noted that Wolf Hollow isn't all too far at all from Portage Township (Cambria County) where a rogue wolf, or was it coyote, known as the Beaver Dam Wolf, was killed in 1907.

Another "spook wolf" was the monstrous white wolf which haunted the area of Sugar Valley (Clinton County). This account is undated, but can likely be placed fairly reliably in the early 1840s; Philip Shreckengast (a secondary account appearing in the *Lock Haven Express* gives his first name as Peter, although Philip, a renowned hunter, is undoubtedly meant) of Tylersville saw the beast emerge from his barn with its mouth smeared with blood. He slammed the door on the beast's tail, severing it, and later used it as decoration on the pommel of his saddle. A short time later, the wolf was seen in the Brungard Cemetery (now known as St. Matthew's Lutheran Evangelical Cemetery) by a travelling preacher, who said that it was yelping "like a yellow cur that had had boiling water poured on it" as it ran out of the consecrated ground.

Jacob Rishel (a *J. Richel* is listed as having owned a farm near Tylersville in 1862) was travelling through Wayne Township towards Lock Haven when he ran into Granny McGill, an old woman widely supposed to be a witch. She told Rishel/Richel that he should try an old hex before contacting George Wilson, who had already slain two supposed werewolves. The good people of Sugar Valley needed to get a hold of a black lamb, born during a new moon in autumn. One was eventually procured, and put in a trap on Tunis Knob (the newspaper account from earlier says *Cooper's* Knob), location of the wolf's lair. The wolf was then trapped and killed.

The head of the wolf was displayed mounted on a pole near Jacob Rishel's home. Children were afraid to pass by the head. It was reported that the jaws of the severed head moved, and that the eyes flashed green, and that as long as it was present it warded away wolves from Rishel's farm.

The pelt of the white wolf was described as being shaggy and long-haired, almost like an Angora goat, rather than short-haired like the average wolf. It is possible that the white wolf was an Arctic wolf ranging far to the south of its normal habitat as this would be an adequate description of their sometimes lengthy fur.

Shoemaker also recounts in his account of the white wolf in "Wolf Days in Pennsylvania" that another wolf which he likened to the French Beast of Gévaudan terrorized the area around Carroll, although no date is given for this event. This wolf was killed by John Schrack, who also had the pelt of the white wolf, as it was attempting a 16-foot leap over a stockade fence at a sheep pen. One of its paws was impaled on the top of the stockade for a good luck charm (he mentions that this tradition was a widespread one in Sugar Valley and elsewhere).

The newspaper article says that while the white wolf was, indeed, real folklore, there were doubts about whether the tale really happened. Shreckengast, for his own part, was the type of larger-than-life figure tall tales swirled around; he once claimed to have choked a bear and scared it from a tree. He also had a somewhat strange/cruel tendency to kill domestic dogs as if they were wild animals.

But if it really was a wayward Arctic wolf, could there have been a population of them in the area, albeit a small one? The likewise larger-than-life King Heizmann (mentioned earlier), mounted on his broken-down old white nag, shot and killed a white wolf (which he lured out with his wolf calls) whose fur answered the description of the Sugar Valley monster, in Union County.

White spook wolves are one thing, but who ever heard of a white spook rabbit? That was just what haunted Buckwampum Hill, in Bucks County. The giant white rabbit could not be caught, and legend said it could only be killed by a silver bullet. Oddly, the area of Rocks State Park, in Harford County, Maryland, is frequented by a nearly-identical being called the "Witch-Rabbit." That lycanthropic lagomorph was supposed to be the transmuted form of a witch, but, identically to the Buckwampum rabbit, could only be slain by a silver bullet.

Sometime between 1800 and 1825, a Polish immigrant by the name of John Wallize was engaged as a wolf hunter in the northern wildernesses, that huge wooded area known as the Black Forest. He hired two Indian guides and another man named Jack Berry. They travelled to the valley of Windfall Run in Potter County; Wallize and the others swiftly constructed a cabin near a waterfall. The hunter claimed to have caught over 130 wolves in the valley in just over two weeks (a number which sounds to be a bit of hunter's exaggeration to me).

A series of snow storms and an avalanche succeeded in completely burying the cabin, leaving only a tiny opening in the chimney for ventilation. Then the wolves began to dig out the cabin. Most likely it was due to their sensing warm-blooded prey inside, although there seems to be an undercurrent of supernatural vengeance for the killing of scores of their brethren, as well. Wallize claimed that "every wolf in Windfall Valley" was engaged in digging. The hunting party chopped down the wall toward which the wolves were making their way; finally, a "pack of lean, hideous looking beasts," supposedly numbering an amazing 300 animals, was seen outside. The party responded to being dug out by killing the wolves. Ungrateful!

## Hans Graf haunts
It was in high school that I first heard the following tale, although I didn't know it at the time. The story as I heard it was that there was a cemetery near Marietta (Lancaster County) with a brick wall and no gate. If you walked backwards around the cemetery on top of the wall thirteen times, the tale went, a "vampire" was summoned and it chased you away. The story always seemed pretty damned unlikely to me, and I disregarded it for years.

The Hans Graf Cemetery, known as the Shock Cemetery in official records (the railroad crossing of Shock's Mills was located just a short way to the west) and which is, according to one historian at the Lancaster County Historical Society, also known as the Wildcat Cemetery, is a burial ground in the tiny little village of Rowenna, located along Route 441 about a mile west of Marietta, just past the Armstrong warehouses. Though a legend emblazoned on the streetward side of the cemetery's stone wall refers to it as 'God's acre,' in reality it is anything but; the cemetery is a tiny affair barely as large as my living room. The entire cemetery is bounded by a crumbling stone wall about three feet tall, but like the version I'd heard there was no gate allowing entry. Probably as a result, the grounds were pretty well overgrown, save one large patch in the northeast corner (I seem to recall a bare patch being mentioned in the tale I heard

The tiny Shock Cemetery, better known as Hans Graf Cemetery.

as being the grave of the person whose ghost appears, but as I can't remember exactly who told me that story, I can't confirm it. If anyone knows, I'd be happy to hear from you).

Like most good haunted spots, there are a few stories about the place. One has it that if you walk around the perimeter seven times by the light of a full moon, you'll meet your doom. This variation of the story is possibly related to an old Danish superstition that a person wishing to make a deal with the Devil should walk the perimeter of a church three times and call out to Old Nick.

The most common of the tales is of a ghostly white dog, sometimes described as a wolf, seen in the cemetery. The dog has even been seen moving between the gravestones, and in addition other ghostly phenomena have been recorded from within the cemetery's walls. I have even heard a tale of a sudden wind which picked up when the grounds were disturbed; I find this particularly intriguing, given the association sometimes made between British phantom hounds and storms. And, of course, the coincidence of a ghostly dog encountered in a cemetery named Shock – which is eerily similar to the name of Shuck and other black dogs, as well as the Shug-monkey – is indeed notable.

The dog is more often encountered, however, not by being seen, but rather by being heard. It is said that quite often, investigators nearing the cemetery are greeted by the barking of a dog, which often serves to stir up the dogs at neighboring homes, which join in. This barking has been recorded by some investigators, and supposedly intensifies if one scales the wall and enters the cemetery. When I myself made my journey to the cemetery, I was, indeed, greeted by a barking as I passed a large hole torn in the western wall. The barking continued as I made my way around the outside of the cemetery, and stopped once I stepped back onto the road. On the rear (north) wall, I found some quite bizarre graffiti – of course, graffiti on a haunted place like this is to be expected, but this appeared similarly to an X and an hourglass, which I later learned were two letters in the Elder Futhark runic alphabet. The runes were *gebo* (gift) and *dagaz* (day) merkstave (inverse), which together refer to the end of a service or contract. Although it very well may not have meant anything, it is quite strange, as its meaning could have been relevant.

I had made the trip accompanied by Mindi and our dog, who was apparently acting up in the car, running back and forth and eventually hiding in the back seat, which was quite strange behavior for her.

Several aficionados of the hobby known as "geocaching" (in which one signs logbooks hidden in various places and which are found with the aid of a GPS device) have photographed a white cat sitting on the top of the wall. Doubtless, this was a *bona fide* stray and not any sort of phantom, but it's definitely eerie given what takes place there!

A man out chopping wood along Kirksmill Road in Nottingham, Lancaster County, heard the grunts and pants of some dog advancing in his direction. He spun to look in that direction, and to his astonishment saw no dog. He saw the ground disturbed by the footsteps of the invisible dog before he fled.

# Chapter Nineteen
## Bigfoot in Pennsylvania

Bigfoot, and related hominid creatures, are by far the most commonly reported of the unexplained events in Pennsylvania; there is scarcely a region of the state from which reports do not originate. Of all the regions of the state, the western counties – particularly along the Chestnut Ridge, extending throughout metropolitan Pittsburgh and up along the Beaver River – account for nearly a half of all reports made. Most of the reports surface from the four counties surrounding the Chestnut Ridge, which is a range of mountains extending northward from the hills and woods of West Virginia. This chapter will be by no means complete – one or more volumes could be written on the phenomenon. Several cases of hominid sightings have already been mentioned in various chapters of this book, and even this chapter will be by no means complete. One or more volumes could be written about Bigfoot in Pennsylvania.

As the title suggests, this chapter deals with sightings of these creatures in the eastern (and northeastern) portion of the state, a region that has spawned a fair number of encounters with this shambling wildman in its own right.

### The "whistling wild boy of the woods"
Susquehanna County is a hilly region, as are most of the counties of the so-called "Northern Tier." Its rural villages and districts are inhabited by rugged hunters and farmers determined to eke out a living, cold winds blowing down from the Pocono Mountains to the south and east and the Allegheny Plateau to the southwest. Of course, the county is a different place today, but this description is likely an apt one of the region in the 1830s. In those days, attacks from mountain lions, wolves, and possibly even straggling bands of Indians were all a very real possibility.

Around 1836 or 1837, a man picking berries somewhere in the farmlands of Bridgewater Township (surrounding the town of Montrose) encountered a creature "having the appearance of a child seven or eight years old though somewhat slimmer and covered entirely with hair, whistling as it approached; the man tried to pursue it, but it speedily ran away." After this sighting, the creature apparently fled to the north into Silver Lake Township, a more ruggedly hilly terrain directly bordering New York state. About two weeks after the berry-picker's encounter, a teenage boy met with the same whistling creature in the forests.

> "He said it looked like a human being, covered with black hair, about the size of his brother, who was six or seven years old. His gun was some little distance off, and he was very much frightened. He, however, got his gun and shot at the animal..."

It ran away, still whistling, and the boy's father told how the boy would "burst out crying" when describing his encounter. This may be a reference to the puzzling physical and mental symptoms with which UFO witnesses and abductees are often afflicted. The same symptoms have been associated with other unexplained phenomena, most notably sightings of the Mothman in West Virginia in 1966, as well.

## Modern gwyllt won't get me to bed

Rumors of wildmen in the vicinity of Lancaster date back to 1858, when it was noted that a hairy man was breaking into barns, drinking from cows (this, by the way, is an extremely common activity for wildmen, and may be a hint of a folkloric origin). Once startled, he threatened the offending parties and then "fled, bounding like a deer."

In the summer of 1871, a number of beastly howls echoed through the Morgantown (Berks County) valley. Following this, a "man presenting the appearance of a huge overgrown bear" was seen traipsing through the hills of the Welsh Mountains, south of the town, in northern Chester County. As one account of the happenings had it, "[a] prevailing superstition amongst the Dutch is that of the 'Hairy man,' the wild man of the mountains, a being without any other clothing than hirsute abundance, who climbs lofty trees, afflicts children and stock, and defies pursuit and capture."

The Welsh Mountain beast, however, didn't escape capture like his prototype described above. As the *Reading Eagle* had it:

> "He was captured and brought to the hotel of Mr. D.K. Plank. When caught he was very nearly in a nude state, having but a few rags hanging to his body. The hair on his head hangs down his back; his face is very nearly covered with long, bushy hair, giving him the appearance of a gorilla more than of a human being. To questions put to him he said he was a native of Ireland, and had lived in the State of Connecticut for a long time. He gave his name as Thomas Foley, and says he has been roaming in the woods for two years. A good suit of clothes was put on him, and he immediately started out for the mountains, tearing his clothes in strips as he moved along. The horses and cattle belonging to farmers along the mountain run and gallop through the fields continually, as if frightened from some unknown cause; the dogs howl and cry as soon as night approaches...At times he is seen on his hands and feet, moving along with the fleetness of a wild tiger...near the residence of Mr. Robert Yocum, a farmer, his horses and cattle refuse to eat, and are constantly running over the fields as

if some demon was after them. He runs through the bushes with the swiftness of a deer, and persons cannot get near him... A few nights since this "What Is It?" made his appearance in the village, on his hands and feet. In a few minutes the villagers were up in arms...He made at the crowd on all fours..."

The mention of the man's being Irish, as well as the location's geographical proximity to the former Welsh Tract, call to mind a feature of Celtic myth. On several occasions, it is recorded that an individual would be affected by a peculiar sort of mental illness upon being defeated in battle, and withdraw into the forests and become known as a *geilt* (in Ireland) or *gwyllt* (in Wales). In fact, it is likely that the wizard Merlin of Arthurian fame was based on one such figure, Myrddin Gwyllt. The story "How Culhwch Won Olwen," in *The Mabinogion*, describes another, named Keledyr; the Irish tale of Suibne Geilt (Wild Sweeney) is the same. This phenomenon is not unlike the Biblical fate that befell King Nebuchadnezzar.

It is indeed ironic that although the creature is explained away as being an Irishman gone feral, the tale also has elements of apparent high strangeness in its account of the reaction of the neighborhood's animals to the presence of the beast-man. It is well-documented that animals, dogs in particular, have extreme reactions to the presence of Bigfoot entities.

In 1874, the Feral Mr. Foley, or some other "wild man," was again seen near Morgantown. But by this time, he seems to have bulked up, as he

> ...[was] nearly seven feet high, and weighs over two hundred and fifty pounds; he walks generally on all fours, is almost covered with hair, gives unearthly yells and makes all kinds of gestures. His hands and feet are double the size of an ordinary man's, and he presents altogether a horrible appearance.
>
> He approaches the cabins of the settlers in the mountains, carries off their pigs and sheep, and with a demoniac laugh, disappears in the dense forests. The brave spirits of the neighborhood go gunning for him, but whenever they come in sight the monster gives a yell and a jump, and before the hunters have time to pull [the] trigger, he is gone.

There is also some other allusion to high strangeness, as the article seems to suggest that at various times of the year, the wildman also appears as a snake or "a wild beast, of unknown species." The article continues that quite recently (August or September of that year) the man had been seen at Parkesburg (Chester County).

I'm not certain whether the tales of these wildmen have any relation to tales of a red-eyed form, seven feet tall, seen further west along the Welsh Mountains in Lancaster County. The form is usually associated with the ghostly legend of Isaac Kohlmann, a powwower who murdered a girl and was well-known as a fiddler.

## Bigfootery on the Union Canal

On September 7, 1985 a sighting of a Bigfoot creature was made in North Annville Township (Lebanon County). The creature seemed to be neckless, with a pointed head. It strode alongside a fence, crossed a road, and moved out of sight into a field. The witnesses noted that it swung its arms as it walked. On the night of the 14th, the same witnesses heard loud shrieks and screeches from a nearby wooded area for a duration of nearly 20 minutes. Investigation turned up a number of 18-inch footprints with a stride length of over four feet. Nearly forty years before, in 1946, another Bigfoot encounter surfaced only a short distance to the east, near Lebanon. In this case, a farmer saw a hairy creature – which later fled – feeding on a dead cow on his property.

Or so it is written in Rick Berry's *Bigfoot on the East Coast*, as well as in a number of other sources as well, most recently in Brad Steiger's *Real Monsters, Gruesome Critters and Beasts from the Darkside*. Based on the further details in Steiger's book, it becomes clear that the incident being referred to did occur in Lebanon in 1946 – but it was Lebanon, *Indiana* and not Lebanon, *Pennsylvania*. The creature, by the way, was one of the ubiquitous black panthers and not a Bigfoot.

Another series of encounters with one or more Bigfoot-type creatures were made in 1975-1976 in rural Londonderry Township, Dauphin County. It started in February of 1975, when residents of a trailer park along Route 230 when several residents complained of bangings, thumpings, and knockings on the walls of their trailers at night. Elizabeth Cahill heard a knocking at her window and opened her front door to see a creature the height of a man, with "smooth-looking" fur standing at the end of her driveway with bent arms and legs. The creature began advancing, and Cahill slammed the door, stating that she "had this feeling that the longer I stood there, the angrier it got" and that its eyes changed. The best she could describe the change in its eyes was that "they seemed to get shinier." When some neighbors arrived at Cahill's trailer with guns and a dog, they found nothing lurking on the premises – in fact, it was noted that although there was snow lying on the ground, there were no tracks visible.

Several other residents reported the shape – one woman was grabbed by a gorilla-like creature and escaped only when she squirmed out of her coat. Another man saw an apish form shambling nonchalantly through the residential area. Some young boys that summer at the baseball fields at Londonderry Elementary School saw a hairy creature walking out of a field to the rear of the school and towards the building. They fled.

Around the same time, a young girl at the trailer park said she saw a "big bear," which she later described as a monkey, trying to hide behind a tree. The creature was about as tall as a man and covered with red-brown hair.

The events died down for several months, resuming in March of the next year. On March 29, at about 11:30 at night, a woman saw a bright light emanating from some adjacent pine woods. At about the same time, she heard a growling sound. Her husband thought she was imagining it, until he himself heard it. He grabbed a gun and vanished into the night, and while he was gone his wife saw a "dark upright shape" shambling towards the wood. The husband, however, had found nothing in his investigation. Perhaps it was not a coincidence that they were the parents of the young girl who had seen the "big bear."

The Pennsylvania State Police confirmed that they were often called to the trailer park to investigate sightings of hairy creatures. It wasn't the State Police seen by several residents, however, who claimed that the military posted a watch at the east and west entrances to the park. It must be noted that this case, which has undeniable overtones of high-strangeness and conspiracy, also occurred not too far from the Union Canal basin (the canal emptied into the Susquehanna River at nearby Middletown). Given my finding of a rather unimpressive stick weaving in the canal bed in Berks County, I could theorize that Bigfoot creatures are using the canal bed as a sort of trail.

## 1920-1921: a prelude to the "Gettysburg gorilla"

For weeks, horrid screams and screeches were heard at night on the farm of Charles Bolig, near Globe Mills in Snyder County. On one occasion, his son Samuel had shot an ape-like beast, which fell and ran off into the forest. But one fateful evening in December, 1920, while Charles was working in the nearby wood, the sounds of the ape were heard very close to the house. Samuel grabbed a revolver which was in the house and made his way outside, followed by his twin sister. The beast was standing beneath an apple tree and leapt at the boy, attacking him –

> "Upon its third appearance the boy Samuel attempted to shoot it with his revolver. He snapped the trigger three times, each time the weapon refusing to discharge its load, with the gorilla only ten feet away from the lad.
>
> The gorilla becoming enraged by the boy's actions assaulted the lad throwing him to the ground and tearing most of his clothing from his body and strangling him almost to death. For two days the boy was confined to his bed. He is now able to be up and around. Claw marks and skin abrasions made by the gorilla, cover the boy's form from face to feet.
>
> Samuel described the animal as having a color between black and brown, a human-like face and large lips. It walked upon its hind legs and was at least, says the boy, seven feet high. When it ran away it moved upon its four legs...
>
> Articles of food disappeared from the cellar of the Bolig home and it is believed that the gorilla in search of something to eat has been making nocturnal visits to the cellar. Apples partially eaten were found just outside the apple bin where a night prowler evidently has been making his regular visits."

Only a few days later, word came that a similar creature was seen by young Chalmer Brannen in Canoe Creek, in Blair County. The boy was separated from the ape by only a thin wire fence. By the time his father, Samuel, retrieved a firearm the monster had shambled off. As in the Globe Mills case, two pieces of meat were taken from the cellar of George Mattern, nearby. It was told how around December 15, a lumber crew near Claysville had a large-footed man walk around barefoot in the snow to create the illusion of the ape-man's passage. Shortly after this incident, the ape again returned to Snyder County, being seen in Middleburg. Mr. and Mrs. Bruce P. Yeager claimed that their car was attacked by a monster "seven feet high [with] arms and legs like telephone poles [which] jumped at their car, missed it and fell sprawling on the road." Apparently, Mr. Yeager didn't appreciate the humor of this slapstick Bigfoot, because he fired two shots – but he was apparently not any more competent at shooting than the ape creature was at attacking. It ran off into the forest.

Soon, word came from Clearfield County of another ape. This one appeared on the night of January 13, 1921, killing and partially eating a calf on a farm near the town of Wallaceton. A posse was raised and gave chase to the beast. During their hunt, a man named Jim Barron encountered the beast, which picked up a club and chased him for a half mile before knocking him prostrate. Personally, I think this encounter smacks more of a human explanation than a Bigfoot one.

## The "Gettysburg gorilla"

Only a week after the Clearfield County gorilla hunt, a "huge gorilla" was seen seated on a rock near Mount Rock (Cumberland County). The ape "arose, stretched itself, and disappeared into a nearby wood" once it realized it was seen. In what passed for skepticism in those days, a Gettysburg man is quoted as saying, "It is evident that some of my Mount Rock friends are seeing more peculiar visions now than they did before the advent of the Eighteenth amendment." The man was referring, of course, to Prohibition, and his opinion that liquor was responsible for the sighting. Who says that viewpoint is anything new?

The next evening January 21, a creature "described by some as a gorilla and by others as a kangaroo" was seen on Snyder's Hill, halfway between York Springs and Idaville in Adams County. It escaped into the South Mountain ridge. The beast took a few days off before its next appearance – and its next was quite an active day.

Early on the morning of January 26, a resident of Waynesboro (Franklin County) named Harry Shindledecker was walking to work past the baseball fields on Memorial Park Drive when he saw a man-sized black monster. Later that day, Henry Needy saw the creature in Rouzerville. It was "crouched in a heap inside the fence. Henry at first took it to be a dog, but when he called to it the animal gave a gurgling bleat, such as he had never heard before." He threw a rock at the creature, but missed, and it fled on all fours towards the woods of the South Mountain. A group of Rouzerville residents took up arms and mounted a hunt for the ape.

Later that evening, William Flohr and Maurice Molesworth were passing through Monterey, a small valley in the midst of the South Mountain, and were just passing the golf course when they came across "…what they thought was a man approaching them on all fours. Thinking it was a sneak thief they stopped and called to him, asking him what he was doing there. The animal came toward them slowly and then rising and coming forward made some gurgling sounds which were not human...the animal was about five feet in height..." Shortly after that, a fellow by the name of Frank Goetz saw the animal, again in the vicinity of Rouzerville.

The next day, it was still in Rouzerville, when Paul Gonder saw it crossing the road. The next day, it was a bit further south, just above the Maryland state line in Pen Mar; the last sighting of the winter came in February, when two men crossing the mountainous ridge saw the monster. Up to this point, it had mainly made its appearances in Franklin County.

The "gorilla" took another hiatus, this time of several months. Its next appearance was in the town of Gettysburg (Adams County) – it put in an appearance on York Street, when a woman saw a four-foot creature which her husband shot at. Footprints were later found leading in the direction of Biglerville, where it later put in another appearance. The next sighting of the beast was made by a motorist named Howard Mitinger, who saw it sitting on a tree stump along the road south of Fort Loudon, in Franklin County.

The hominid creatures haunting the area returned in July, 1961. Around 10:00 on the night of July 1, a thirteen year-old in a wooded area behind the Greenmount Fire Hall (about six miles south of Gettysburg, just south of where the Marsh Creek crosses the Emmitsburg Road) encountered a long-haired creature about five feet tall, smelling somewhat like manure; in the early 1970s, there were rumors of "werewolves" seen in the South Mountain. I'm not certain that these weren't references to the sightings of the dog-like 'Dwayyo' further south in Frederick County, Maryland, though, because the only sighting of which I'm aware was of a man-sized bipedal creature, which leapt across the roadway in the vicinity of Mount Holly Springs, in Cumberland County.

In the late 1990s, an impressive encounter was reported by a hunter near Pine Ridge Lane in the Michaux State Forest. In late October, 1997, a man hunting deer heard a bizarre howling vocalization, which was followed by a stampede of several deer from the treeline nearby. This was followed by the sighting of a brown-furred, seven or eight foot tall bipedal figure. After a few moments, this one was followed by a smaller individual. A garbage-like odor permeated the air. Was this sighting of a male and his mate?

In January, 1998, a Bigfoot creature was seen walking through a field and stepping over a fence along the Fairfield Road by a man and his wife. The animal was seven or eight feet tall, with self-luminous red eyes. The sighting was interesting in that while the man felt the creature was dark with fairly short hair, his wife felt that it was reddish with long hair.

On Valentine's Day, 2002, a series of simian footprints were found leading along a creekbed near the Waynesboro Reservoir (Adams County). Rick Fisher of the Pennsylvania Bigfoot Society as well as the BFRO investigated the footprints, most of which were a foot or so long, with a relatively short stride-length compared to most Bigfoot sightings. Although a hoax was not proven, one is suspected by some researchers. More tracks were found in 2003 and 2004 (both of these were of isolated tracks or pairs of tracks, suggesting the possibility of some animal brachiating). As no ape was ever sighted, the case remains unresolved.

Bizarrely, an upwelling of water in the Michaux State Forest is referred to as Hairy Spring. The derivation of the name is unknown, but it is tempting to wonder whether the spring was frequented by the apish beasts seen to the south!

## The Suscon Screamer

Just outside Wilkes-Barre, in Luzerne County, is the village of Suscon. The town is home to one of the more famous urban legends of that corner of northeastern Pennsylvania, the so-called "Suscon Screamer." This is some sort of screaming thing reputed to haunt the area around where the Susquehanna Railroad once crossed over Suscon Road south of Wilkes-Barre. The former bridge is also known among locals as the *Boo-Boo Bridge* or, more ominously, the *Black Bridge*.

Unearthly screams have been heard reverberating through the forests near the little town of Suscon for generations and some residents have even phoned the Pittston Township police to complain of the shrieks. Some versions of the story, in traditional ghost story fashion, have it that a ghostly female haunts the area, whether it be a victim of a car crash, a love-crossed suicide, or one of the ubiquitous phantom hitch-hikers. One of the more popular versions has it that the tiny town is haunted by a porcine swamp monster that emerged from one of the surrounding bogs. In the 1970s, the *Wilkes-Barre Times-Leader* reported that a local hunter heard something tramping around through the trees. Through his binoculars, he saw something

> "...about 6' long with a long snout. It weighed about 200 pounds and was gray in color. It had webbed feet with long claws and had a huge head...the ground was clawed up as if 100 turkeys had gone through."

This sighting was actually investigated by the Pennsylvania Game Commission, although the hunter refused to take the investigators to the area due to fear of the monster. The hunter did, however, say that the creature he saw was neither bear nor coyote.

Another popular version has it that the Screamer was actually a panther that escaped from a circus train; although the specific date of this supposed crash is unknown, older residents of Suscon still remembered it, at least as of 1995 (when *Pocono Ghosts, Legends and Lore* by Charles Adams III and David Seibold was published). If this identity of the Screamer were true, by this late date it would doubtless be the cat's restless phantom.

It is also possible that the stories of the Screamer are distorted references to encounters with Bigfoot, whose penchant for blood-curdling shrieks, yells and howls is well-known. Indeed, there was a sighting of an apparent family group of four man-sized sasquatch made in 1976 near a lake just south of Suscon. The Wilkes-Barre/Scranton Valley has produced a number of reports of Bigfoot creatures, some as long ago as early December, 1906 when an "enormous ape" which scaled trees and shied away from men was seen near Georgetown.

2003 saw a sighting from Dickson City, a town on the ridge overlooking Scranton (Lackawanna County), when two hunters saw a seven-foot being covered in chestnut-brown hair. In the summer of 2008, a sighting was made of a four-foot white hominid creature in Carbondale, at the northern end of the valley. The creature was seen loitering about in a wooded area bordering on territory that had once been a strip mine.

The southwestern portion of the state (better known as the Chestnut Ridge, after a range of wooded mountains extending northward from West Virginia) is home central for the Pennsylvania Bigfoot Society, and that's probably not a coincidence. After all, about half of all the Bigfoot sightings reported in the state surface from just two counties, Westmoreland and Fayette – although, bizarrely, quite a number have been reported from the city of Pittsburgh itself. In fact, in his 2003 book *Bigfoot! The True Story of Apes in America*, cryptozoologist Loren Coleman (ever fond of lists) named this selfsame region one of the "Twenty Best Places to See a Bigfoot."

## Modern gwyllt, part two

The Uniontown (Fayette County) *Morning Herald* and *Daily News Standard* reported throughout 1929 on an early story of a wildman – or wild*men*. On December 9, 1928, a farmer named Hilary Wilburn told his wife he was going to hunt bear and that he would be back by noon. Wilburn didn't turn up at the appointed time, and over the next few weeks a controversy began. There were apparently encounters with a wildman near that town, which was at first held to be Wilburn – this even though his wife declared that Wilburn was dead and that the beast was not he.

On January 3, a rifle – later determined to have belonged to the missing man – and some clothing was found at the home of J.L. Reams in Confluence, just over the Somerset County border. Local farmers reported that their hogs were being stolen, and furthermore, "persons who have seen the 'wild man' claim that when they approach he takes to his heels and runs like a deer." On January 7, a bag containing a cooked hog's tail was found at the home of William Lewis a short distance away in Ursina. A fire was found to have been lit on top of the woodstove – this despite the fact that a fire was already burning *in* the stove. There were even rumors that the culprit was found "in an attempt at skinning a pig alive." No footprints could be found in the snows near the cabin, and the newspaper made the tongue-in-cheek suggestion that "maybe the cave man has taken to the trees and nests and eats with the birdies."

A few days after the finds at the Lewis cabin, a girl named Lena Henning, again at Ursina, claimed that she was accosted and chased by the wildman. Shortly after Henning was pursued, a young boy named Edward Gower claimed that a man was skulking around in some rocks on his father's farm. A group of men formed a search party, and soon found chicken bones and a knife in a clearing among the boulders – but, incredibly, no sign of a fire, suggesting that the chicken was eaten raw. Nearby, they claimed to have found a black man, "of medium height, thick beard and somber expression," asleep on a log. They pursued him through foot-deep snow, but he escaped.

The last appearance of the wildman (rather, of a trace of him) came in the latter part of January, when the remains of an (old) fire and chicken bones were discovered near Ohiopyle. Footprints were found in the snow, and they were followed nearly 500 yards until they abruptly stopped at a high cliff.

Here the story gets more bizarre. It was around this time that a man named William Moore wandered into some town or another in Somerset County. He claimed that he had gotten lost in the mountains, and had wandered through the wilderness for 43 days (slightly beating Moses' record). He was admitted to the county home in Somerset, apparently suffering from exposure. There were also rumors that Moore's mental state was being questioned. There were rumors for a time that he was the wildman seen near Confluence – but these were later disproven, and he died. What's odd is that once Moore turned up, the black man pursued there was no longer reported.

There the sightings died – but all was brought full circle when, in August of 1929, the body of Hilary Wilburn was found only a half mile from his house. It was noted that the body was being taken away for autopsy, but that's the last it was mentioned. If it did, indeed, turn out that Wilburn had been dead the entire time, not only should some suspicion be cast on his wife, who apparently knew his fate, but we're left with the question of exactly *who* these wildmen were. The early sightings make no mention of his having been black, so I'd assume that we're dealing with at least *two* separate wildmen here.

I've found, by the way, that it was quite common for wildman reports in the 1920s and 1930s to be blamed on blacks – presumably, just another facet of the rampant racism of the era.

In 1976, a book entitled *The Creature: Personal Experiences with Bigfoot* appeared in print. Under the pseudonym "Jan Klement," a local geology professor told of experiences he had had at his secluded cabin some four years before with a creature he christened "Kong." After a number of sightings, Klement managed to lure the creature closer to his home with gifts of apples (presumably a favored food source of Bigfoot creatures, as many instances are on record of the man-apes devouring the fruit). In his book, Klement stated that the beast was nearly 7 feet tall, and was far more human than ape, a hairy, primitive-looking man with a large stomach.

The book went on to describe how Kong killed and ate deer, rabbits, and other animals – and later went into extensive detail on such topics as the length of Bigfoot's penis, and at least this individual's apparent proclivity toward sex with cows.

After a few months (in January 1973), Klement came upon the dead body of Kong and buried the body of the Sasquatch in the Sandy Creek-Winds Gap area of southern Fayette County. He said that when he returned to the burial site a few months later, the body couldn't be found – it had apparently been exhumed.

The mystery of Klement's true identity has never been resolved. There were rumors that it was Paul Johnson, a researcher into the hairy hominids of western Pennsylvania, but those rumors proved to be unfounded. Later, rumors surfaced that Klement was actually John Tomikel, an earth science teacher from Cuddy, but it was determined that wasn't so, either. Tomikel said that the *real* author had been dead for a decade (as of 2009). Whoever 'Jan Klement' was, we are left with the fact that these encounters weren't the beginning of the sightings along Chestnut Ridge. They did, however, provide a convenient segue into the next phase of sightings, which began only a few months later.

Throughout the summer and autumn of 1973, the sightings of Bigfoot creatures came flooding in from Westmoreland and Fayette Counties. The names Derry, Latrobe, Greensburg, Uniontown, and Ligonier turn up again and again as hotspots of activity. Westmoreland seemed to be the more active of the two counties, with no less than fourteen sightings recorded from the month of August alone. Multiple sightings were occurring daily, and the sightings were widely spread apart, making it much less likely that these were of one exceptionally mobile entity.

Perhaps 'entity' is the right word – because many of these sightings occurred in a disturbingly close proximity to UFO activity, or contained some other element of definite strangeness (I'll deal with these strange Bigfoot in a cursory way, since an excellent book entitled *Silent Invasion* has appeared chronicling just this phenomenon, written by Stan Gordon, the original investigator of many of these events). One such was a bipedal creature which appeared in Fayette County in November, 1973. One of those stereotypical trigger-happy witnesses shot the Bigfoot, which promptly disappeared. It later reappeared, was promptly shot again, and this time vanished.

Some of the weirder Bigfoot reports seem to be associated with mists in some way. An old example of this type of seemingly paranormal Bigfoot was made in around 1912 near Gallipolis, Ohio when the two witnesses reported that they were followed by a "dark cloud" previous to the appearance of what is very much a classic Bigfoot entity.

Two of these bizarre foggy Sasquatch were seen in Fayette County, at Jumonville Summit near Uniontown (site of a battle in which Joseph Coulon de Villiers, Sieur de Jumonville, was slain during the French-Indian War). In two separate instances occurring in 1973 and 1975, Bigfoot creatures were encountered that vanished as a fog rolled in. In the latter of the two, the creature apparently "floated" rather than walked. This detail is reminiscent of a sighting that occurred on September 22, 1980, in Trufford, Westmoreland County. The figure was described only as "tall" and was a shiny black in color. He, too, floated rather than walked.

Another was seen in late July of 1994, when in Cambria County, Joe Nemanich and Tom Johnson were driving on Route 422 between the cleverly-named towns of Revloc and Colver. As they crested a hill, they came across something striding across the road in front of their truck. They were about 300 feet away from a barrel-chested dark brown humanoid with a large upper body. Nemanich said that as the beast ran up an embankment at the roadside, he noticed a "type of mist" covering the creature's feet, and when they went back to the sighting area they found that the Bigfoot – or is it Mistfoot – hadn't left any tracks.

Back on the night of July 31, 1966, Betty Klem and Douglas Tibbets were sitting in their stranded car on Beach 6 at Presque Isle State Park (Erie County) when they saw a star-shaped UFO coming out of the air. When two policemen nearby flipped on their red siren light, the craft vanished. But apparently it had left something behind. Tibbets and the two policemen began walking towards another part of the beach, but before they had gone very far Klem began frantically blowing the car horn. She had seen a "dark, featureless creature walking in front of the car. It seemed to have the general shape of an upright, large creature, but was not any kind of animal she had seen before..."

The incident at Presque Isle, incidentally, is one of the 700+ reports of UFOs that the U.S. Air Force could not explain away during their investigations. They discounted the sighting of the monster seen by Miss Klem, but since when is the military concerned with shadowy monsters, anyway?

I've seen this incident cited as a Bigfoot encounter, although I guess the monster seen by Miss Klem could have just as easily been some other sort of alien – it *is* described rather vaguely. I suppose in retrospect it could have been a bipedal bear seen quickly, but that smacks of skepticism, wouldn't you say? And besides, we'd still be left with the weird coincidence of exactly why a bear would be at a UFO landing site.

But one of the classic – the classic, in fact – report of Bigfoot of a possibly alien kind surfaced on the night of October 25, 1973 from the tiny little village of Leisenring, near Uniontown (in fact, this sighting is often recorded as having occurred *in* Uniontown, although it didn't). The details of the case are positively harrowing. Several people saw a reddish sphere descending towards a field, and the farmer who owned the land dispatched his son to investigate. Two local boys tagged along. As the son (who used the name "Stephen" in the initial report) neared the field in question, the headlights of his car began to dim.

Stephen and the boys saw a large white dome sitting in the field, luminous. Harsh sounds – sounds like a lawnmower – emanated from the object. Screaming sounds were heard all around, and a burning-tire like odor hung in the air. Two large beings – 7 feet tall and 8 feet tall – were advancing along the fence. They had glowing green eyes and were a dirty gray color. The humanoid beings had a "neckless" appearance, and their legs appeared stiff. They seemed to communicate with a plaintive, baby-like whine. Stephen fired his gun at one of the beings, but it did not slow. Again he fired, and again the beings plodded on. Finally, he fired three shots directly into one of the humanoids, which made a whining noise. The dome vanished, and the creatures turned and walked back into the forest.

A short time later, the State police were contacted and they went to the field in the company of Stephen to find a glowing area – which was shunned by livestock – in the place where the dome once sat. Other bizarre happenings were recorded after Stan Gordon and the other investigators arrived on the scene. Stephen went into some sort of fit in the field, jerking around and making animal-type sounds. He muttered, "it's in the corner." (This sounds like some sort of epileptic reaction to me.)

## Shenango Valley shenanigans

Linda Godfrey reports that she had received word from Ohio researcher Jason Van Hoose of a monster called the Shenango Valley Werewolf,

> four to five feet tall, covered with patches of long, black hair, a piglike nose, large round 'fisheyes,' a disturbing mouthful of snaggle teeth, and perhaps the strangest feature of all, its elbow and knee joints bend opposite of a human, somewhat like a dog but not really. It has been described as a mutation between a man and a dog. The face is flat without a snout. It can run on all fours and is also bipedal. Some witnesses stated that it moves very quickly, almost to the point of seeming to appear or vanish in thin air. When it runs on all fours, the elbows jut forward and the hands, which should be angled in to the body, are turned outward. Its hands are similar to a human's, with digits, but when running the fingers are clenched, giving the appearance of paws. There is no tail.

The Shenango Valley (the region of Mercer County), though, has long been a hotbed of Bigfoot reports. From that same vicinity comes the legend of the so-called Pig People, beings which are supposed to be ghosts connected with a leper colony which once existed in the area. Could conceptions of Bigfoot and the Pig People have merged and the conflated tales have given rise to another, more zooform, monstrosity?

On the night of March 13, 1989, Diana Baroni of Wheatland reported that her dog was exceedingly restless. Thinking the dog wanted out, she opened the door; it ran out, turned around and came right back in. Wondering what it was that was prowling around their house, she called for her husband, Chris. He went back outside with the dog, which led him to the edge of a wooded area. After a few moments, it ran off in pursuit of whatever was running through the woods and being none too quiet about it. Mr. Baroni called the dog to return, and made his way back into the house, where Diana said she had been hearing noises.

After a time, the two, accompanied by the dog, made their way back outside and found a number of footprints – 11 inches long and 5½ wide, and five-toed. Whatever the creature was, it had nearly a four-foot stride. The Baronis called the Pennsylvania Game Commission, who found a wide trail of broken trees in the forest where something had passed through.

A sad case was reported in 1909 by the *New Castle News*. Constable William Hallis and Herbert Reardon of the South Sharon police department were executing a search warrant at the home of a Mr. and Mrs. Mathews in Wheatland when they discovered, in a tiny upstairs room,

> a strange, half-naked creature, so vile and filthy that words are inadequate to describe him. The few shreds of clothing revealed a hairy creature, with resemblance to a human being, so dirty and wretched in appearance that the officer and his companion instinctively shrunk back in alarm...His body was covered with a long growth of hair like an animal, with long, bushy eyebrows at least two inches

> in length, and a beard that extended below the waist...it was learned by questioning that he is a son of the woman by a former marriage...

This sad tale is concluded with the note that "Physicians who examined the prisoner this afternoon say that he is demented, but harmless, and that with proper care and attention he would today be in fairly good shape. It is feared, however, that the long years of confinement and neglect have totally destroyed his reason. He will be sent to an asylum."

# Chapter Twenty
## The Rest of the Weird

As I was compiling the reports and stories mentioned in this book, I of course came across a number of stories which were incredibly odd and just didn't fit anywhere else. These are the true zooforms, those pesky creatures seen once and then (usually) never again. Some are of a more or less humanoid nature and seem to be a sort of phantom, though none can say exactly who the phantom is of. A few were called aliens in the original reports, although that's seemingly only because the witnesses really didn't know what else to call it, as there's often no associated UFO phenomena present.

### The Green Man

One of the most interesting – if grisly – phantoms featuring in Pennsylvania lore is the Green Man, haunting roads near Pittsburgh. The story varies as do most urban legends, but usually goes like this: a worker on the high-tension electric wires had a horrible accident, surprisingly one not caused directly by the wires. Maybe he was struck by lightning. That varies too. Either way, he was electrocuted.

His skin became green, the flesh melted from his face, and a hole was burnt in his cheek. Any number of tunnels around Pittsburgh are said to be the site of whatever accident befell the poor unfortunate and the usual haunts of his cigarette-smoking, green-glowing self. If you turn off your car's headlights in the 'Green Man Tunnel,' he will come close to your car. Wonderful! No sight I'd rather see than a faceless phantom chomping on a smoke. But there's a drawback, though, because all of that electricity stored up in his body will discharge and overload your car battery, causing you to be stranded in Pittsburgh at night, probably in the wee hours of the morning. Not high up on my list of things to do. Obviously, being stranded at night in Pittsburgh *in the immediate vicinity of a walking Tesla Coil* is even further down my list.

Probably the most gruesome thing about this tale is that it was true in part. Ray Robinson was eight years old in 1918, when he was badly electrocuted on a railroad bridge near Beaver Falls (Beaver County). A full 1,200 volts coursed through his body. The accident scarred him badly, and as a result he lost both eyes, his nose, an ear, and an arm.

Soon, stories spread throughout the area of a character called 'Charlie No-Face' and people lined Route 351 at night to look for the spectre, man, nobody knew what he was. Of course, it was none other than Ray Robinson, who, because of his disfigurement, only left his house at night. People would ply him with beer and cigarettes, occasionally posing for photos, and by all accounts Ray was a genial fellow. In 1985, at the age of 75, Ray died, but the story of the Green Man survives until today.

The green's a mystery though. Some say Ray's skin did get a greenish tinge to it, others say it was reflective clothing he wore to avoid being hit by a car.

Another entity of an electrical nature, this one more nebulous, put in an appearance in July, 1970. Five young people at a backyard swimming pool in Chester (Delaware County) saw a huge, man-shaped figure seemingly made out of living electricity step out of a nearby telephone pole and quickly run into the house beside the swimming pool. At the moment it vanished into the house all the fuses were blown. These witnesses very well could have merely witnessed an explosive power surge and attributed humanoid characteristics to it. Alternately, the surge of electricity could have altered the perceptions of the witnesses. As any ghost hunter will tell you, surges of electromagnetic energy are believed to alter the perceptions of people. I, for one, would be interested to know whether the path along which the 'man' ran happened to be aligned with the power line suspended above.

## Kids see the damnedest things

As should probably come as no surprise, some of the oddest reports come from out of the mouths of babes and as should likewise come as no surprise, good deal of them can be dismissed as fantasy or misinterpretation. On the afternoon of September 24, 1996 some students at an elementary school in northern Indiana County (two 9-year olds) saw a 4-foot blue creature in a cornfield near their school. The overgrown Smurf had a sort of antenna on its head, its face was shaped like a cone, and at the terminus of the cone it had one eye. Groovy. Its gait was similar to a soldier marching.

A famous incident in the annals of Pennsylvania's weird lore took place in March of 1981 by a storm drain in New Kensington (Westmoreland County). A 3-foot tall, crested, green creature, roughly humanoid and tailed and referred to as a "baby dinosaur" by the young witnesses was seen standing by the drain. It squealed when one of the children lunged forward and grabbed it.

In a 2000 article, Chad Arment writes that the being was likely a basilisk lizard, whose brilliant green color, pronounced crest and occasionally bipedal stance are well-known. Elsewhere in this volume, I have noted that a small alligator had previously been seen in New Kensington. Alligators, though, are obviously completely quadrupedal and lack any sort of crest but still, this sighting should not be necessarily ruled out as a possible alligator sighting. I would also forward the possibility that the being could have been based on the Dave Sutherland drawing of the Troglodyte appearing in the *Advanced Dungeons & Dragons Monster Manual*, which had appeared in 1978.

## Weird "aliens"

On August 3, 2000, five people in a field near Uniontown (Fayette County) saw a man-sized being with skinny arms and legs floating along a few inches above the ground. The being made no sounds at all, and the witnesses also commented that birds and crickets were silent as it passed by. The floating man also carried a brown "staff," and vanished as the witnesses moved to get a closer look.

The only thing better than some kind of extraterrestrial wizard, is a spacewalking purple giant. And, surprise! Just such an event was recorded on the night of April 15, 1973. A small flap of UFO sightings was occurring around Manor (Westmoreland County). During this flap, a man was driving along the Penn Manor Road, near a power station just outside the town, when he saw a white-lit object descending over some railroad tracks. Oddest of all, a gigantic figure standing anywhere from 8-10 feet in height was floating at the end of a line, like an astronaut on a spacewalk. As the witness watched, this new recruit of the Grape Ape Space Academy blinked out of existence like "a string of lights."

A fellow in Philadelphia in 2010, apparently not content to see *one* strange animal, went for the threesome – get your collective minds out of the collective gutters. I mean he saw *three* separate creatures, all totally different! First was a thunderbird, then a giant mantis, then a little clanking robot – but I'm not going to subject anyone to the full, mind-numbing details of his sighting, which were written in a bizarre, rambling style. Clearly this poor Philadelphian was either baked out of his gourd on some hallucinogen, or just plain drunk off his ass.

## Goatmen

Ruscombmanor Township, in the northern Oley Hills directly south of Fleetwood (the same region where a wallaby – origin still undetermined – was seen in 2007), Berks County, was host to an unusual creature in October, 1879. A man named Schmehl, a son of the prison inspector, said that he saw an object lying near the entrance to a gated field, and when he approached, it stood up and ran at him. It was something out of a nightmare: only four feet tall, with hairless yellow skin and only two clawed fingers on each hand. To top it all off, the thing apparently had two goat-like horns. The beast ran at Mr. Schmehl, and then he and a local man named Jared Rissmiller went in pursuit of it. They found the apparently lazy dwarfish monster again lying on the ground. It stood up and ran off into a forest. Another local, named Mr. Heckman, felt that it may have been an ape of some kind.

Another goatman creature was reported from the Big Valley of Lancaster County sometime in 1973. The two Amish brothers who were attacked by the monster said it was "the size of a good heifer, gray in color with a white mane. It had tiger-like fangs and curved horns like a billy goat, ran upright on long legs and had long grizzly claws." The following day, another man was working in his fields when the monster began running at him. He took up a scythe, but the goatish figure grabbed the scythe and crunched it between its jaws, devouring the wooden handle. Apparently, the creature's goatish nature extended to its diet, as well! The night after that, a woman claimed that the startled creature hurled a dead goose at her as it fled into the woods. No information on whether it just picked up a dead goose or whether it had killed the unfortunate animal.

I tried unsuccessfully to place this encounter for years, as there is no locale in Lancaster County of which I'm aware called the Big Valley. Through discussion with Chad Arment, he relayed to me his feeling that possibly SITU got it wrong and the encounters were actually placed to the north, in Mifflin County. That county *does* have a region called Big Valley (actually Belleville), which, by way of making things more complicated than they at first seem, is merely a small section of a larger valley called the Kishacoquillas. And to further the relationship of the two, the area around Belleville is inhabited by Amish farmers who emigrated from Lancaster County. Did SITU see a reference to the Amish and Pennsylvania and assume it was in Lancaster?

## Bagunk and the Blur

Animated mists and fogs are an eerie, but unexplained, phenomenon. Well-known from Great Britain, they also have an undeniable connection of some sort with humanoid creatures – both Bigfoot and apparent sightings of what are termed ghosts. However, the phenomenon is also recorded quite often from Pennsylvania, and is particularly prominent in the mountainous areas of the north and west.

Perhaps the most famous, and one to which I alluded earlier, is the entity called "Bagunk" which haunts two cemeteries (St. Michael's and St. Andrew's) along Main Street near Glen Lyon, a small town in Luzerne County between Shickshinny and Nanticoke. It's a frustrating ghost, since nobody can say with any certainty from which of the two graveyards it emanates. While investigating the tale, paranormal author Charles J. Adams III spoke with an elderly gentleman from the area who claimed to have had an encounter with the Bagunk. He said that this happened when he was just a boy and that even then, parents were telling their children to stay away from the cemeteries "because of the monster" – so clearly, the legend has existed for a while.

In one of the two graveyards, the old man said it was St. Michael's, a sudden tempest kicked up (like one of the tales I've heard about Hans Graf Cemetery in Rowenna) as he and some other teenagers were prowling about. The children saw "a cloudy figure gliding just over the grass. It had the rough shape of a human being, but didn't make any motions...[it moved] through a couple of tombstones. It acted like it had no idea we were there, and didn't much care if we were."

The arts building on the campus of Lock Haven University in Clinton County, Sloan Hall, is haunted by a trio of entities. One is a child's ghost reputed to haunt the third floor (I myself have encountered many an odd feeling while attending classes there); one is a White Lady, as has been described in a previous chapter; but the most notorious is one called the Black Blur. The specter is an eerie form, sometimes said to be simply a cloud of thick smoke which moves with its own intelligence and deliberation. It exudes a feeling of malevolence and malice, and it is reputed that on one occasion, the shadowy form pursued two students through the backstage areas of Sloan's main downstairs theater.

One of the more mysterious aspects of Sloan Hall's phantasmagoria is a supposed antagonism existing between the white and black beings. One student said the smoky cloud rushed at the

stage in the theater, but the White Lady appeared and the malevolent apparition stopped short at the edge of the stage. On another occasion, in the late 1980s, another student heard an unexplained sound, and then the Blur began to flit around, inspiring feelings of malaise and general sickness. There was a flash of white light – and the Blur was dispelled.

Although the "White Lady vs. the Black Blur" aspect is easily dismissed as urban legend among college students, the presence of the dark ghost is not, particularly as it sounds as if it is similar to the fast-moving black "shadow people" noted from haunted structures worldwide. Another possible explanation, however, is found nearby. On the hillside behind Sloan Hall, and extending westward for a fifth of a mile is a huge cemetery (well, cemeteries, as like the Bagunk's haunts they're properly two separate burying grounds which run together). This cemetery is reputedly frequented by gigantic, black, ghostly dogs, as well as being frequented by Satanists and occultists in rumor. Could Sloan Hall's vaporous entity be related to these dogs in some way?

It isn't the first time that black dogs, noted in British legend as shapeshifters, have appeared as misty forms. A phantasmal hound haunts the lanes near Brereton, a village on the monster-haunted Cannock Chase, and in the mid-1980s a woman living on the Chase offered an encounter of her own. At about 11:30 at night, she and her husband were driving along Coal Pit Lane when, as she said in a letter appearing in the *Cannock Advertiser*:

> "...the headlights picked out a misty shape which moved across the road and into the trees opposite.
>
> We both saw it. It had no definite shape seeming to be a ribbon of mist about 18 inches to 2 feet in depth and perhaps 9 or 10 feet long with a definite beginning and end. It was a clear, warm night with no mist anywhere else. We were both rather stunned and my husband's first words were: "My goodness! Did you see that?""

While there is no definitive evidence that this misty form was directly related to the phantom hound, a connection is indeed tempting.

Another interesting aspect of this case, in reference to the Bagunk, is the name of the lane where the encounter took place: *Coal Pit*. The area of northern Pennsylvania from which these misty encounters have surfaced is well-known for its rich veins of coal, and Glen Lyon in particular definitely *was* the site of a coal mine. Is it possible that animate mist is somehow connected with the presence of coal?

## The Warren Wendigo, storms, and spirits

Pennsylvanians experienced heavy snowfall in the winter of 2009-2010. There's something undeniably eerie and otherworldly about the dead of night just after a heavy snowfall. The sky has a bizarre color and the wind howls through the trees... it was just such a night when a truly monstrous entity appeared somewhere in Warren County. In a series of e-mails received by Pastor Robin Swope, a person naming themselves only Jacob described an encounter he had had in December, 2009 just after a heavy snowfall.

Early on the morning of December 19, Jacob had risen early to tend to his family's chicken coop. As he approached the barn, he heard a horrible shrieking from somewhere nearby: "like a wounded animal crying at the top of its lungs but with a man's voice. All I could think of was someone stuck out in the storm, so I walked out to the field to see where the screaming was coming from." As he made his way across the field, the snow still falling around his head, Jacob paused. Ahead of him, barely visible through the driving winter storm, stood a dark shape, "like a giant, almost twice the size of a man, but with a man's shape. It was either covered in fur or wore an animal fur coat because I could see the hair moving in the wind. As I got to it, the screaming became louder and louder, and from the sight of it and the sound of it I knew that this thing was just wrong. I had to get away from it as fast as possible. I ran as fast as I could and made sure the screaming thing was behind me. After a few minutes the snow was not coming down as hard and I found myself back in the field behind the shed. It was then that the screaming voice had stopped."

What was this monstrous being? One of the hairy man-beasts reported quite often in this general area of the state? Or was it something else, something more... primal? The tribes of Native Americans throughout the United States and Canada have tales of monstrous beings they called wendigo, usually perceived as the remnants of men who turned to cannibalism, but in earlier traditions it is seen as a transmuted sorcerer, immune to cold; in some tales, he either vomits ice or his heart is made of ice. The Canadian tribesmen are known, at times, to suffer from a peculiar type of insanity in which the afflicted believes himself possessed by a Wendigo and embarks on a cannibalistic spree. And the Inuit of Alaska know a similar malady, called by them *piblokto*, in which an individual sprints into a raging snowstorm, disrobed and mimicking the cries of animals and birds. Though he is less malevolent in nature than his northern cousins, even the Pueblo tribe of New Mexico have tales of a spirit named Shakak who wore robes of icicles and could use snow and hail as weapons; he battled with, and was conquered by, the spirit of summer.

Henry W. Shoemaker, in his 1920 book *North Mountain Mementos*, recounts earlier stories of similar winter-dwelling wildmen. Around the time of the Civil War (1861-1865), a young man who lived near Benton was having an illicit love affair with the wife of an enlisted man. During a blizzard one February he was driving his sleigh towards the woman's home when he spied a "huge glittering stalagmite icicle by the edge of the trail" near a lumber camp. Shoemaker goes on: "As he approached he could see that it was a huge, white-bearded man, glistening like silver, though there was no moon, and even if there had been, its rays could not have shown through the dense forest which overhung the road. When he came abreast of it, the icy monster leaped into the cutter beside him, throwing its long, cold arms about the rich man's neck, squeezing the breath out of his throat in a vice-like grasp."

The young man eventually escaped the snowman, and finally arrived, weak and out of breath, at the cabin of his mistress. He collapsed, and his illicit lover sent him to bed. Three days later, the man died in some sort of fit – it was as if he were once more locked in combat with the icy assassin. His features were twisted and horrible upon his death.

The undertaker who brought the young man's body out of the mountaintop cabin stopped at the lumber camp where the wildman had first been seen, and there he saw "very large footprints, bigger than could have been made by the corpse with his wild cat moccasins."

Some lumbermen working on the so-called Dogback Trail between Brush Valley and Lock Haven saw a wildman roaming in the forest. They didn't attempt to communicate with the wanderer or, indeed, even hesitate a moment before one of the men shot the poor unfortunate. The hermit later froze to death; thereafter, on moonlit nights, the wildman's ghost arose from the icy waters and sat on the teamsters' sledge, rendering it immobile. Try as the woodsmen might to shoo it away, the ghost would not move until daybreak. Eventually, the man who shot the wildman was found dead, his features (as the story recounted above) distorted and twisted. Shoemaker also mentions two other tales of wildmen – one, called the Wildman of the Storm, appeared to lumbermen along Mosquito Creek in Clearfield County, warned them of a blizzard, and vanished; the other, unnamed, appeared along Trout Creek eating a raw rabbit.

Here I should refer to the Centre County tale of the spook wolves of Elk Creek Gap, which came off the slopes of Hundrick Mountain to alight on the sledges and wagons of lumbermen. If a spook wolf was sitting on the wagon – even if it only had its paw on it – the wagon was rendered immobile. Some of the superstitious teamsters drew or painted hex signs on their wagons, but all for nought. Eventually, the spooks moved on. The similarity of this tale to the activities of the Wildman of the Dogback Trail are notable, especially since the site of these eerie happenings were on the other side of Brush Valley only a mile or so to the south.

Lake Erie has long been notorious for its famously bad weather and the frequency with which ships go missing, leading some to posit a Great Lakes Triangle in competition with the famous one in Bermuda (in all fairness, the idea of a Great Lakes Triangle is probably unnecessary – I'm quite certain any ships missing there are due to the weather). These phenomena have led sailors to tell stories of another storm-spirit. The spirit's usually seen as a female, of yellow eye and skin tinged sea-green. Her hands are bedecked with talons, and her arms are long; she's sometimes called Jenny Greenteeth, in a not-so-subtle allusion to British legend, but more often simply the Storm Hag. Like a mythological siren, she sings a song:

> *Come into the water, love,*
> *Dance beneath the waves,*
> *Where dwell the bones of sailor lads*
> *Inside my saffron caves.*

The stories vary as to whether she actually causes the storms or whether she just lies in wait, sinking the vessel just when the sailors think they've weathered the storm. This certainly was the case in one account of her manifestation, in 1782. A ship was sailing from Canada, approaching Erie, when a storm began. It died down as they neared the shore, but then the malevolent hag sprung up out of the water, sinking the ship (I haven't managed to find any record of this event, though – although it may be a reference to the British schooner *Beaver*, which ran aground near the mouth of the Cuyahoga River, near Cleveland, Ohio, in 1786). At any rate, the screams of the dying sailors still emanate from the lake.

I should here also mention that, aside from the previously mentioned hags like Jenny Greenteeth, witches who could influence the weather at sea – whether for boon or bane – were an element of naval folklore worldwide, but particularly in Britain, presumably since Britain historically has a strongly maritime culture. Many of these sea witches still feature in the folklore of the New England states. The Lake Erie area, and Fort Presque Isle in particular, was a focal point for the British military of the 1700s battling their French enemies to the north and west.

## The Mold Man of Pittsburgh

One of the more bizarre tales of an ostensibly ghostly nature – though quite possibly its nature is otherwise – that I am aware of comes from the Pittsburgh vicinity, and is recalled only as having taken place sometime in the 1970s. A number of bodies were being moved from cheaply-built crypts into a cemetery's new mausoleum. One of the corpses exhumed was that of an elderly gentleman, which was quite well-preserved. The body was covered in a fuzzy layer of greenish mold. Nonetheless, the stoic undertakers interred the body in its new resting place.

In the morning, when the first employees trickled in, they found that the door to the mausoleum hung open – some substance smeared across it – and one of the individual crypts was open as well, and empty to boot. The panicked employees ran to fetch a telephone to call the police, who would doubtless come to the cemetery and be confronted by a rather morbid case of grave robbery. Or would they? For when the employees passed the old, poor-quality crypts, they found one was open and contained a coffin.

The mold-covered corpse was in the coffin, of course. The smear on the door was some of the man's fungal covering, and there were bits of mold deposited on the floor of the mausoleum as well. The body was dutifully returned to its new resting place, but a short time later the same thing happened, with one new detail – a trail of footprints leading through the muddy earth to the old crypt. This time, however, the caretakers wised up and got a priest to give the dead fellow his Last Rites again; which must have worked, since the fungal zombie stayed in his new "home" this time.

## The spectral stickmen
In the winter of 2002, investigator Rick Fisher encountered some sort of creature in Lancaster County, near the intersection of Marietta Pike (Route 23) and Kinderhook Road outside of Columbia. The figure he saw was nearly five feet tall, was furry, black, and quite skinny (Rick estimated it was probably no more than 70 pounds). Being an investigator with the Pennsylvania Bigfoot Society, he at first thought the beast was a juvenile Bigfoot-type creature, possibly a severely malnourished one, but later dismissed the possibility. If it was that malnourished, it was rather unlikely it would be just striding along, but would probably be more likely to be off somewhere dying. The thing turned its head and glared at him with yellowish eyes as he focused his car's headlights on it. After it looked at him defiantly – as if to say "go ahead, try it" – it vanished. It didn't run off the road – it just faded from view. A few months later, he heard from someone else who had seen it near the same junction – and then he also heard the story of Dwight, who saw a similar skinny hairy guy in 2000 about two miles from where Rick's encounter took place. Interestingly, the tiny village of Kinderhook is also mentioned in a *Frederick* (Maryland) *News* article appearing on November 2, 1912. Although that article is focused on some dogs killed in Marietta, it mentions, tantalizingly, that "some time ago, a wild man was seen at Kinder Hook, some miles east." However, at present I have no further information on the Kinderhook Wildman.

I'm still actively looking for stories from newspapers and such (finding more information for the Kinderhook Wildman is one focus), and I've found a good deal since I wrote this book, so perhaps a more direct sequel will be done someday (as is, I'm working on a follow-up, focused on Maryland. Of course, I'm naturally not as familiar with Maryland as I am with Pennsylvania, but I have a decent bit – if you'd be interested in that?). What exactly it was is yet to be determined, but in 2009 it – or something similar – was encountered in Cumberland County. The witness was at an apartment complex in Mechanicsburg when he heard noises in a tree. He looked towards the tree, and saw an odd creature descending.

> It appeared to be approx. 3 – 3½ feet in height. As it seemed to be running fast across the lawn, it sounded heavy, like an adult human. It didn't have fur or feathers. I saw two skinny bone like legs, but didn't look as though it had feet like a human, [but they were] similar to that of a bird.

Oddest of all, the creature appeared to be *transparent*.

(It must be said that the witness provided photographs of this being's footprints, which resembled those of a deer; a deer seen head-on rather than from the side could appear similar to the creature seen.)

## When taxidermy walks

The works of Pennsylvania folklorist Henry W. Shoemaker contain several tales attributing a sort of animating force or at least supernatural qualities to even the preserved remains of animals. One of the most famous is the tale of the "spook wolves", a number of wolves mounted by a Czech taxidermist by the name of Jake Zerkow. The wolves were originally displayed at the Philadelphia Centennial; somehow, they ended up in a stone house in Potter County. From there, the wolves' bodies were said to futilely prowl the woods nearby, unable to make any kills.

A mountain lion was killed by a hunter in 1864 near Centerville (Snyder County). The hunter, strangely, mounted it on the roof of his house; at some point, the lion's mate leapt onto the roof and carried the mounted specimen into the forests of Jack's Mountain, where it returned to life. Troxelville (also in Snyder County) has a similar tale in which a mountain lion's skin stalks the forests of the White Mountains.

On Christmas Eve, 1868 hunter Lewis Dorman shot a panther on Shreiner Mountain, again in Snyder County. After he had it mounted, he donated the mount to a museum in New Berlin, where it was rumored to leave its glass case and prowl the building, hunting and killing mice. The taxidermed mount has since been transferred to Albright College.

Adam Kriegbaum, a farmer, killed a pair of golden eagles on his farm near Muncy (Lycoming County). When he returned to his homestead the next day after a night of drinking to celebrate his kills, he fell ill and his crops were planted late; Kriegbaum lost an entire season of crops as a result. One night there was a terrible thunderstorm, during which the window the stuffed birds were kept near broke. What seemed to be a large, black-winged form entered the room and enveloped the mounted eagles; they were gone. Furthermore, since that time no eagles have returned to the area. The story of Kriegbaum and his eagles seems to have some overtones of the Thunderbird myth; an association between the avian form and storms is noted. Unfortunately, the account contained in Shoemaker's *North Mountain Mementos* does not give a date for the tale.

## Mysteries of the worm

Unsurprisingly, the next Pennsylvania tale comes from Westmoreland County, at the end of the Bigfoot-haunted – and otherwise Fortean-haunted – Chestnut Ridge.

From the files of researcher Stan Gordon comes the story of a group of motorists, one of whom wrote to him on June 27, 2009 telling of how, near Mount Pleasant, they saw a creature which "looked like a giant caterpillar." This creature was about 7 inches long and was tapered at the ends. Its body looked to be segmented, and it was a bluish-white color although the

**The almost freakishly large Hickory Horned Devil caterpillar.**

witness noted that it was not luminous. The sighting of this creature would be strange enough, but over the next few moments, the witness reported, the four motorists saw no less than six or seven of the gigantic worms.

That story was bizarre enough, but in September of that year it was to get even stranger as investigator Brian Seech, of the Center for Unexplained Events, forwarded to Stan a second report of the giant caterpillars – and, if possible, this encounter was even stranger than the first.

The witnesses in this instance, near Youngwood, saw what they took to be a snake, four feet in length, which slithered into the road and before their eyes "broke apart" into eight smaller creatures between six and eight inches in length. Each of these smaller creatures was dark in color, with a "shiny wet" texture. These animals assumed a ring-like position, following which they reformed themselves into the form of the snake and moved away again.

Perhaps the oddest thing about these reports, then, is that they may not be odd at all!

The caterpillars of the Royal walnut moth, also called the Regal moth (*Citheronia regalis*), the so-called "Hickory Horned Devils," can grow to seven inches in length – moreover, they are

## The Mystery Animals of Pennsylvania

of a pale green coloration and a segmented, chitinous appearance. The caterpillars are more commonly found in Ohio than in Pennsylvania, but there are some records of them from the state. They have large reddish horn-like appendages which give them their name, but these are not always visible.

The second sighting also has an explanation. A number of varieties of caterpillar migrate in a processional pattern, one behind the other. From a distance, this procession would resemble a snake. It is even more interesting that it was noted that the components assumed a ring-like shape before moving on; Hickory Horned Devils are not known to have processionary habits, however.

## Tall-tale critters

Finally, we conclude this journey into the truly bizarre with some tales that aren't too serious. Folklorists have collected and written bestiaries detailing the odd creatures in which the lumberjacks working in woods throughout the United States believed.

A few of these supposedly inhabited Pennsylvania – given the prevalence of the lumber industry in the state's history, this isn't really much of a surprise.

BALL-TAILED CAT

The first is a little arboreal beastie called a Ball-tailed Cat (*Felis caudaglobosa*), said to inhabit only certain areas of Oregon and Sullivan County in this state.

*Fearsome Critters*, by Henry H. Tryon (1939) describes it thus:

"Its chief physical characteristic is a hardy heavy, bony ball on the

## The Mystery Animals of Pennsylvania

end of its tail. The feet are clawed as with all true cats, making it an excellent climber; and this species has the stealthy habit of lying out on a limb, and when the unsuspecting lumberjack passes beneath, the Cat drops on its victim and pounds him to death with the ball. In the rutting season the male uses this instrument to call the female by drumming on a hollow log.

This species has occasioned much discussion and peppery argument.

SQUONK

It has often been confused with both the Sliver Cat and the Dimaul. A careful Study of the equipment and habits of the three species shows plainly that they are, by no means the same. It is quite possible that they are all distantly related; perhaps the Ball-tailed boy is a less highly developed variant of the same phylum."

The same book describes another creature, the Squonk (*Lacrimacorpus dissolvens*), a supposed resident of Pennsylvania's northern forests.

"Probably the homeliest animal in the world, and knows it. The distribution was once fairly wide, the usual habitat being high plains where desert vegetation was abundant. History shows beyond dispute that, as these areas gradually changed to swampy, lake-dotted country the Squonk was forced to take to the water. Of distinctly low mentality it traveled constantly around the unaccustomed marches in search of fodder. With time, it developed webbing between its toes, but only on the submerged left feet. Hence, on entering the water it could swim only in circles, and never got back to shore. Fossil bones dredged from these lake-bottoms reveal that thousands perished of starvation in this manner.

To-day the Squonk is met with solely in the hemlock forests of Pennsylvania. It is a most retiring bashful, crepuscular animal, garbed in a loose, warty, singular ill-fitting skin. The Squonk is always unhappy—even morbid. He is given to constant weeping over his really upsetting appearance, and can sometimes be tracked by his tear-stained trail. Moonlight nights are best for Squonk hunts, for then the animal

## The Mystery Animals of Pennsylvania

of a pale green coloration and a segmented, chitinous appearance. The caterpillars are more commonly found in Ohio than in Pennsylvania, but there are some records of them from the state. They have large reddish horn-like appendages which give them their name, but these are not always visible.

The second sighting also has an explanation. A number of varieties of caterpillar migrate in a processional pattern, one behind the other. From a distance, this procession would resemble a snake. It is even more interesting that it was noted that the components assumed a ring-like shape before moving on; Hickory Horned Devils are not known to have processionary habits, however.

### Tall-tale critters

Finally, we conclude this journey into the truly bizarre with some tales that aren't too serious. Folklorists have collected and written bestiaries detailing the odd creatures in which the lumberjacks working in woods throughout the United States believed.

A few of these supposedly inhabited Pennsylvania – given the prevalence of the lumber industry in the state's history, this isn't really much of a surprise.

BALL-TAILED CAT

The first is a little arboreal beastie called a Ball-tailed Cat (*Felis caudaglobosa*), said to inhabit only certain areas of Oregon and Sullivan County in this state.

*Fearsome Critters*, by Henry H. Tryon (1939) describes it thus:

"Its chief physical characteristic is a hardy heavy, bony ball on the end of its tail. The feet are clawed as with all true cats, making it an excellent climber; and this species has the stealthy habit of lying out on a limb, and when the unsuspecting lumberjack passes beneath, the Cat drops on its victim and pounds him to death with the ball. In the rutting season the male uses this instrument to call the female by drumming on a hollow log.

This species has occasioned much discussion and peppery argument. It has often been confused with both the Sliver Cat and the Dimaul. A careful Study of the equipment and habits of the three species shows plainly that they are, by no means the same. It is quite possible that they are all distantly related; perhaps the Ball-tailed boy is a less highly developed variant of the same phylum."

The same book describes another creature, the Squonk (*Lacrimacorpus dissolvens*), a supposed resident of Pennsylvania's northern forests.

SQUONK

"Probably the homeliest animal in the world, and knows it. The distribution was once fairly wide, the usual habitat being high plains where desert vegetation was abundant. History shows beyond dispute that, as these areas gradually changed to swampy, lake-dotted country the Squonk was forced to take to the water. Of distinctly low mentality it traveled constantly around the unaccustomed marches in search of fodder. With time, it developed webbing between its toes, but only on the submerged left feet. Hence, on entering the water it could swim only in circles, and never got back to shore. Fossil bones dredged from these lake-bottoms reveal that thousands perished of starvation in this manner.

To-day the Squonk is met with solely in the hemlock forests of Pennsylvania. It is a most retiring bashful, crepuscular animal, garbed in a loose, warty, singular ill-fitting skin. The Squonk is always unhappy—even morbid. He is given to constant weeping over his really upsetting appearance, and can sometimes be tracked by his tear-stained trail. Moonlight nights are best for Squonk hunts, for then the animal

# APPENDIX
The Great Circus Train Wreck of 1893
(from the *Huntingdon Daily News*, May 29, 1943)

Inevitably, in discussions of out-of-place and otherwise cryptozoologically relevant happenings (particularly in older accounts), the explanation of the animal having been an escapee from a passing circus train, zoo, or some other sort of menagerie is commonly proposed. I wish to use this appendix to offer, (mostly) word for word, an extremely thorough news story I uncovered about just one of these circus train wrecks, by way of showing that they are not all just "cop-out" explanations, that they do, indeed, happen. The story contains a few cryptozoologically-relevant sightings; and this circus wreck should be considered as a possible explanation for any erratic animals seen around that time.

### 1893 Opening In Geneva
He opened the season of 1893, on April 22, at Geneva, his home town, as well as winter quarters. The event always brought crowds from many distant Ohio points, and the town was gaily decorated for this occasion which taxed the capacity of the "big top" by those who gathered there in wishing their popular young townsman success on his tour for the season. The next stop was at Youngstown, Ohio, and the show swung into Pennsylvania for Beaver Falls. On through western and south-western Pennsylvania it continued until May 18, when it was back in northeastern Ohio for Painesville. From this area it again moved into Pennsylvania, playing at Franklin on May 19, then heading southeast, eventually reached Houtzdale on May 29.

### Fateful Descent
On completing its engagement at Houtzdale the show was delayed for some unknown reason in leaving that place for its next day's performance at Lewistown, a distance of near 82 miles, and it was 2:15 a.m. before the wheels began moving. At Osceola further delay was encountered; this time apparently in obtaining motive power. The train finally got under way and passed Summit at 5:09 a.m. beginning its fateful descent down the mountain. The crew with engine 1590, was as follows: Steve Creswell, engineer; Harry Meis, fireman; Wil-

liam Snyder, conductor; James Barger, flagman; with William Heverly, John Grazier and Harry Myers as brakemen. The train consisted of 17 cars and caboose, with the car next to engine carrying the elephants 'Jennie' and 'Lizzie,' while the last three cars were sleepers, with performers and other show members. Other equipment of the train consisted of several long flat cars, upon which the wagon cages loaded with animals were placed, as well as other wagons with show equipment. A number of stock cars were also carried, loaded with horses, ponies, and other animals.

## What Happened?

What happened from this point on has always been a subject of much discussion. It must be understood that at this time the air brake was in its infancy; the rolling stock of our railroads not all being equipped with this safety appliance until a later date, when its installation was made compulsory by enactment of Federal and State laws. Some have claimed that the train had gained such momentum that the engineer found he had insufficient braking power to check the alarming rate of speed his train had attained. While members of the show later told of being frightened by the speed of the train, the train crew denied that it had gotten out of control.

## Car Topples Over

Engineer Creswell stated that as approaching McCann's crossing he noticed a slight jump of the train when something about the first car seemed to break, at the same time tearing the tender loose from the engine. Instantly the first car appeared to topple over, with the following cars leaving the rails, all occurring in a few flashing moments to the accompaniment of a thunderous crash of splintered wood and twisted steel. For a fleeting instant, which seemed magnified many times, a death-like silence seemed to enfold the scene, to be broken by a terrified uproar from the injured and panic-stricken animals.

Before the sun had set on this beautiful day in late May, the greatest catastrophe ever to have happened in American circus history had been recorded.

William LaRue, member of an acrobatic trio, who was riding in one of the sleepers, wrote to his parents in Philadelphia that same day:

"I jumped off the train before it stopped and fell down a small embankment. When I got up, there stood a tiger in front of me and on the other side were three lions, all loose. I did not linger long out of the car [*neither would I*]. Had the sleepers gone 20 feet further we all would have been killed. And how the poor dumb brutes did suffer! One elephant had its leg broken, horses were ground up, people cried like children, and the poor working men were all cut and mangled. Blood was seen along the track for three miles, where the horses had walked."

Of the cars involved in this crash, some were totally demolished, while the three rear sleepers were miraculously prevented from being included; the occupants suffering minor bruises or being shaken up, although Mr. Main had received a slight injury to his neck from which he still continues to feel the effects.

## Five Men Killed

The lives of five men were lost, these unfortunate being: William Haverly, brakeman, of Tyrone; Frank Traine [*ironic*], ticket agent, of Indianapolis (who was sleeping in the ticket wagon at the time), Thomas Lee and Barney Multaney, canvasmen, and John Strayer, of Houtzdale, who had boarded the circus train at that place and was erroneously reported as being a Main employee. Eleven injured were taken to the Altoona hospital, some in a critical condition, and despite reported deaths of some, all appear to have recovered.

## Killed 2 Days Later

Robert M. Gates, aged 28, employed on the work train, was killed two days later while aiding in clearing the wreck, when struck by a suddenly broken cable.

Of a total of 127 horses, colts and ponies, 69 were either killed or had to be killed. The toll of trained horses was exceptionally high. Sixteen wagon cages which contained menagerie animals were also destroyed; the cages in a manner offering some protection to their inhabitants. The crash caused many to be broken open, giving freedom to the beasts and reptiles.

An incident of the disaster occurred at the car which contained the elephants. One of the animals was found to have its head pinned down by the wreckage and hardly had it been removed when the heavy brute struggled to its feet, shaking off the large timbers like straw. 'Jennie,' the smaller of the two elephants, suffered an injury to its foreleg, and although not broken, as the previously mentioned performer had claimed, it was permanently lamed to a certain extent. 'Lizzie,' the other elephant, was uninjured.

## Vivid Recollections

Dr. W. Frank Beck, of Altoona, has vivid recollections of the Main disaster. Residing in Tyrone at the time, he was returning from Altoona that morning and stepping from the train at Tyrone an atmosphere of excitement was immediately noted. Asking the reason, he was informed of the wreck, at the same time observing a number of carriages and buggies driving away with those who had hurried to the scene.

Dr. Beck was also taken to the disaster and recalls many incidents to which he was a witness. Some snakes which had not escaped to the wood were being recaptured by show attaches. The lions appeared as somewhat cowed, but the elephants were in a somewhat different mood; pulling up turf and throwing it high in the air, as though to show their disapproval of the calamity. Many monkeys had escaped to nearby trees where they kept up an incessant chatter. Several lady performers were observed sadly petting the injured horses, at the same time sobbing bitterly.

Dr. Beck also tells of seeing the dead employees being taken from the wreckage as well as the finding of one some distance away, who had doubtless wandered there when overtaken by death.

Dr. Gemmill, the surgeon for the Pennsylvania Railroad in Tyrone at the time, whom Dr. Beck found at the wreck upon his arrival, also aided in caring for the injured.

Gen. Daniel Hastings, elected governor the following year, awaited the departure at Tyrone of the train for Clearfield that morning, where he was to give an address for Memorial Day, which was doubtless prevented by the wreck.

### Owner In Last Sleeper

Of all accounts, either oral or written, few have ever mentioned what Mr. Main had been doing immediately before or after the disaster. As we sat in the den on the second floor of his comfortable home, this question was asked: "Where were you riding at the time of the wreck and how would you describe the scene of destruction as you first beheld it?"

### Tiger Attacks

"I was riding in the last sleeper," he stated, "and of course we were all in sleeping apparel at the time. After leaving the car I could not believe that the scene before me was other than a dream. I recall seeing a tiger spring for the neck of a zebra, which escaped with some injury and later recovered. The tiger then attacked and killed a sacred cow, later killing a cow belonging to a farmer, nearby, so I was told. While much confusion and uproar came from the various animals, all our people were, as a rule, very calm.

"We learned that Mr. Traine was pinned among the wreckage, but as many of the freed animals were gathered around this spot none would venture in attempting a rescue. My mother determined to reach the unfortunate man, and securing a broom, proceeded to swing it right and left as she went forward, clearing a path through the dazed brutes which blocked the way. She succeeded in reaching the dying man; none of the animals attempting to molest her."

Traine realized that his last moments were at hand and gave instructions that his body be sent to his widowed mother in Indianapolis. As rescuers lifted the beam which held him prisoner, death immediately ensued.

## Bitten By Lion

Of the most ferocious beasts which escaped, all were either killed or captured before serious harm could be visited upon the residents of the locality. Two show employees met injury in being bitten by a lion; these beasts, however, were captured without much difficulty, one being taken by a trainer who securely tied it to a tree with a heavy rope. Another lion was apparently choked to death by its struggles in resisting attempts to drag it to confinement.

The tigers proved to be the most vicious and were eventually killed. Another dangerous animal was a huge ape which presented a problem in its capture. This creature moved but a short distance from the scene and took possession of a stump where it continued to sit for many hours. It did not attempt to harm any person or escape, but bared its teeth to any that came too close.

## Amusing Capture Of Ape

Among the employees familiar with the animals was a Negro, who some believed could aid in the capture by getting a rope over its head, which led Mr. Main to say:

"Jasper, go over and see what you can do; he won't hurt you."

"I knows he won't, Mr. Main," said Jasper, "'cause I ain't gwine tech 'im!"

The ape's capture was successfully accomplished when two parties proceeded with lassoes at the same time from front and rear, getting both ropes over its head.

What was the cause of the disaster as the train reached McCann's crossing? Many are inclined to believe that the shifting of the load in the elephant car as it reached the reverse curve caused it to topple over and leave the rails. Mr. Main is rather inclined to share this theory. The jury selected by Coroner Poet gave as their verdict on June 5: "The cause was the fast running of the train down the mountain of the Clearfield branch, Tyrone division, Pennsylvania Railroad."

The majority of the animals which were not killed made no attempt to escape, but roamed around content with their short freedom and did not go far from the wreck. These included a water buffalo, two camels, a dromedary, zebra, yak, hyena and many small animals from various parts of the world. Alligators were stretched upon the ground as if dead, but a rub along the nose with a stick proved them wide awake.

The animals which escaped, as well as those supposed to have escaped, have furnished the basis for countless tales and these in turn have been somewhat elaborated with passing years. Many stories were thrilling indeed, of animals which continued to live in the Alleghenies, but those who related them never explained how certain jungle beasts became so readily acclimated to severe Pennsylvania winters.

## Farmer Kills Tiger

The tiger which killed the sacred ox soon appeared at the farm of Alfred Thomas, not far from the accident, where a woman was milking a cow. The frightened woman hastily disappeared as the beast sprang upon the cow, killing it. The farmer soon appeared with a rifle and put an end to the tiger.

The next report of an escaped animal came in a dispatch from Shamokin, and told of a panther appearing the day following the wreck, some distance from Mt. Carmel and confronting a man driving along the highway in a two-horse wagon, who was carrying the U.S. Mail to a distant point. Curiously enough, the driver's name was Frank buck, but he showed no intention to 'bring it back alive.'

The animal was apparently being tracked by two dogs, which appeared on the scene as Buck fired his revolver at the beast. The panther, springing into the wagon, was likewise followed by the dogs, whereupon Buck, having missed the panther with his first shot, fired again and this time was believed to have hit his mark. A short battle in the wagon proved enough for all concerned, and when the panther leaped out and slunk away into the bushes, the dogs, as well as the mail carrier, had no inclination to follow. The dispatch closed with this curious explanation of the beast's appearance:

"Strange as it seems, many think the panther, after the circus wreck at Tyrone, jumped into an open box car standing near. A freight train from that place reached Shamokin early Wednesday morning, and the beast is supposed to have ridden all the way. Being used to traveling in cars it would be quite natural for it to seek such shelter. It is about 190 miles from Tyrone to the place where Buck had his thrilling combat."

## Animals Killed By Farmers

Another animal, reported as being a black tiger, was claimed to have been killed by a farmer at Bald Eagle on June 2. This beast was supposed to have been in pursuit of some sheep at the time. On the following day a 'silver tip panther' was also claimed to have been shot by Gotleib Wagner at his farm a mile north of McCann's crossing, as it was making a meal of his chickens, at 2 o'clock that morning.

The *Tyrone Herald* on June 10, carried this item:

"Another of Walter L. Main's wild animals was killed yesterday by John Parker and Robert Snyder, near Vail. The latter was fishing near where the wreck occurred, when a wild hog made a dash for him. Dropping his fishing rod he immediately picked up a gun and fired, killing the beast."

*Kangaroo Seen On Warrior Ridge*
Perhaps the most unusual story of escaped animals came from Huntingdon county, when early in June numerous travelers who crossed Warrior Ridge between Huntingdon and Alexandria scarcely believed their eyes when they saw a kangaroo hopping through the woods. How this animal managed to get so far away from it natural habitat in the antipodes was believed solved by the Main disaster. It later disappeared from this section and was next reported as having been seen in Mifflin County. However, five years later, in the summer of 1898, John P. Swope, among the last of old Pennsylvania trappers, reported seeing a new species of animal on Warrior Ridge, while making his rounds in gathering pelts. The aged trapper described this animal: "Of the kangaroo order, with nose like a sheep, large eyes, short fore legs and long hind legs. When startled, stands upright on hind legs."

Mr. Main had no accurate record of the loss of animals and recalling the preceding stories to him, he stated that the animal mentioned as a black tiger was doubtless a black panther. He had seen this animal escaping to the mountains and had often wondered what its fate had been. The report of its killing at Bald Eagle, as related to him by the writer, was therefore the first knowledge he had of it. The number of panthers reported seen and killed in this and surrounding counties shortly after the disaster was remarkable indeed. The official records in Harrisburg would indicate that the last bounty claim paid on a panther was in 1886, and yet the description of many of these animals could not be identified with any belonging to the circus. Those reported were black, white, or spotted, which coloring was entirely unlike the native Pennsylvania lion or panther, which was of a brown or tawny shade."

# BIBLIOGRAPHY & ACKNOWLEDGMENTS

First, I would like to acknowledge the following newspapers, both active and inactive, which provided articles used in the composition of this book. All are from Pennsylvania unless otherwise indicated: *Adams County Sentinel, Altoona Mirror, Beaver County Times, Bucks County Gazette, Chester Times, Chillicothe Constitution-Tribune* (Missouri), *Clearfield Progress, Cumberland Evening Times* (Maryland), *Frederick Daily News* (Maryland), *Greenville Evening Record, Hagerstown Daily Mail* (Maryland), *Harrisburg Patriot-News, Huntingdon Journal, Indiana Evening Gazette, Indiana Progress, Lancaster Examiner and Herald, Lancaster Intelligencer, Lebanon Daily News, Lehigh Register, Lock Haven Express, Lowell Sun* (Massachusetts), *Marietta Register, McKean County Democrat, Middletown Daily Argus* (New York), *Mobile Register* (Alabama), *Monessen Daily Independent, New Castle News, New Holland Clarion, New York Times, Ogdensburg Journal* (New York), *Oswego Commercial Times* (New York), *Philadelphia Evening Bulletin, Pittsburgh Post-Gazette, Pittsburgh Press, Pocono Record, Pottsville Republican, Reading Eagle, Sioux County Herald* (Iowa), *The Call, The Morning Call, Titusville Herald, Towanda Daily Review, Uniontown Daily News Standard* and *Morning Herald, Wayne Independent, Wellsboro Gazette, Williamsport Gazette and Bulletin, Wisconsin State Journal* (Wisconsin).

First and foremost, I should acknowledge my grandfather, Albert Weisser. Whether he knows it or not, his stories of unusual animals seen in Potter County were a big influence on the formation of this book. I should also acknowledge my father, Donald Gable, and other members of his family whose fascination with history and the supernatural helped shape this book as well. Loren Coleman, in addition to many other books, wrote *Mysterious America*, which was an early influence and still stands out in my mind as the first real cryptozoological book I read. The title of this book is a tribute to the debt I owe him. In addition, throughout various stages of writing, several individuals have helped by offering opinions, references, or otherwise information of which I was previously unaware: in alphabetical order, the ones I can think of immediately are Chad Arment, Jon Downes, George Eberhart, Rick Fisher, Jeffrey R. Frazier, Linda Godfrey, John Moore, Darren Naish, Richard Nagle, Theo Paijmans, Nick Redfern, Ben S. Roesch, and Karl Shuker. I'm pretty certain I've forgotten one or more people somewhere along the line, so don't take offense if you're not on this list.

The following books were also of use:

- Adams, Charles J. *Ghost Stories of Berks County, Book One* (Exeter House, 1982)
-----. *Pennsylvania Dutch Country Ghosts, Legends and Lore* (Exeter House, 1994)
-----. *Berks the Bizarre* (Exeter House, 1995)
-----. *Montgomery County Ghost Stories* (Exeter House, 2000)
-----. *Ghost Stories of Chester County and the Brandywine Valley* (Exeter House, 2001)
-----. *Ghost Stories of Berks County, Book Two* (Exeter House, 2002 ed.)
-----. *Coal Country Ghosts, Legends and Lore* (Exeter House, 2004)
-----. *Luzerne and Lackawanna Counties Ghosts, Legends and Lore* (Exeter House, 2007)
- Albertson, Charles L. *History of Waverly, New York* (Waverly Sun, 1943)
- Allen, Joel Asaph. *History of North American Pinnipeds* (Government Printing Office, 1880)
- Andren, Kari. "Black bear roaming around Upper Allen Twp." *Harrisburg Patriot-News* (May 9, 2010)
- Arment, Chad. "Giant Amerindians." *North American Bio-Fortean Review* 1 (April 1999)
-----. "Dinos in the U.S.A.: a summary of North American bipedal 'lizard' reports." *North American Bio-Fortean Review* 4 (2000)
-----. "More odd 'wildcat' reports." *North American Bio-Fortean Review* 4 (2000)
-----. "Giant snakes in Pennsylvania." *North American Bio-Fortean Review* 5 (December 2000)
-----. *The Historical Bigfoot* (Coachwhip, 2006)
-----. *Boss Snakes: Stories and Sightings of Giant Snakes in North America* (Coachwhip, 2008)
-----. *Varmints: Mystery Carnivores of North America* (Coachwhip, 2010)
- Arnold, Neil and Mathijs Kroon. "Mystery animals of the Netherlands: part one." *Cryptozoology Online (blog).* http://forteanzoology.blogspot.com/2010/11/neil-arnold-and-mathijs-kroon-mystery.html
- Ashton, John. *Curious Creatures in Zoology* (John C. Nimmo, 1890)
- Baring-Gould, Sabine. *The Book of Werewolves* (Cosimo, 2008)
- Becker, Marshall J. "The stature of a Susquehannock population of the mid-16th Century based on skeletal remains from 46HM73." *Pennsylvania Archaeologist* 61:2 (September 1991)
- Bierhorst, John. *Mythology of the Lenape: Guide and Texts* (University of Arizona Press, 1995)
- Bleech, Mike. "Looking for a state-record fish?" *Altoona Mirror* (February 24, 2003)
- Bord, Janet and Colin. *Bigfoot Casebook Updated* (Pine Winds, 2006)
- Brathwaite, D.H. "Notes on the weight flying ability, habitat, and prey of Haast's eagle." *Notornis* 39 (1992)
- Buck, William J. *Local Sketches and Legends Pertaining to Bucks and Montgomery Counties* (1887)
- Butcher, Scott D. and Dinah Roseberry. *Spooky York, Pennsylvania* (Schiffer, 2008)
- Cadzow, Donald. *Archaeological Studies of the Susquehannock Indians of Pennsylvania* (Pennsylvania Historical Commission, 1936)
- Chase, Maxwell. "The public record." *Pensacola Independent News* (July 10, 2010):

- http://inweekly.net/article.asp?artID=11605
- Chorvinsky, Mark. "Phantom dogs in Maryland." *Strange Magazine* 19 (1998)
- Chorvinsky, Mark and Mark Opsasnick. "A field guide to the monsters and mystery animals of Maryland." *Strange Magazine* 5 (1989)
- Clark, Jerome. *Unexplained!* 2nd Ed. (Visible Ink, 1999)
- Cohan, Jeffrey. "Binturong moved to out-of-sight digs at zoo." *Pittsburgh Post-Gazette* (April 24, 2002)
- -----. "Breeding days are numbered for Binny the Asian bearcat." *Pittsburgh Post-Gazette* (August 5, 2002)
- Coleman, Loren. "Windigo: being some remarks on the Native encounters with and traditions of the many-monikered hairy hominid of Eastern North America, also known as the Marked Hominid and the Eastern Bigfoot." *CRYPTO Hominology Special I* (2001).
- -----. *Mysterious America: The Revised Edition* (Paraview, 2001)
- -----. *Bigfoot! The True Story of Apes in America* (Paraview/Pocket Books, 2003)
- -----. "Mysterious 'Jan Klement' story resurfaces." *Cryptomundo* (March 26, 2009): http://www.cryptomundo.com/cryptozoo-news/art-encounter/
- Coleman, Loren and Bruce G. Hallenbeck. *Monsters of New Jersey* (Stackpole Books, 2010)
- Coleman, Loren and Patrick Huyghe. *The Field Guide to Bigfoot, Yeti and Other Mystery Primates Worldwide* (Avon, 1999)
- Connor, Matt. "Specters of Sloan: apparitions abound at LHU arts center." *Lock Haven Express* (October 24, 2009)
- Courogen, Chris A. "Bear in Upper Allen Township struck by car, but police think it's still on the loose." *Harrisburg Patriot-News* (May 10, 2010)
- -----. "Upper Allen Township's wayward bear headed back to woods following capture." *Harrisburg Patriot-News* (May 11, 2010)
- -----. "Black bear spotted in North Annville Township, Lebanon County." *Harrisburg Patriot-News* (May 25, 2010)
- -----. "Bear captured in Lemoyne park is latest found in area suburbs." *Harrisburg Patriot-News* (May 27, 2010)
- -----. "Yet another bear captured in Harrisburg's West Shore suburbs." *Harrisburg Patriot-News* (June 8, 2010)
- Cox, Joseph. "Shore lines." *Lock Haven Express* (July 18, 1969)
- Cox, William T. *Fearsome Creatures of the Lumberwoods* (Judd & Detweiler Inc., 1910)
- Crable, Ad. "Our very own Hogzilla." *Lancaster New Era* (August 11, 2004)
- Curtin, Jeremiah. *Seneca Indian Myths* (E.P. Dutton, 1922)
- Dahlgren, Madeleine Vinton. *South Mountain Magic* (Lethe Press, 2002)
- Davis, William W.H. *The History of Bucks County, Pennsylvania* (Democrat Book and Job Office Printers, 1876)
- Durant, P.A. *The History of Cumberland and Adams Counties, Pennsylvania* (Warner & Beers, 1886)

- Durham, Michelle. "Live alligators being caught in the wilds of Warminster, Bucks County." http://www.kyw1060.com/Another-Live-Alligator-Caught-in-Warminster--Pa-/2643384.
- Elias, Joe. "Hampden Township police warn residents about black bear." *Harrisburg Patriot-News* (May 24, 2010)
- Engels, Jeremy. "Equipped for murder: the Paxton Boys and 'the spirit of killing all Indians' in Pennsylvania, 1763-1764." *Rhetoric & Public Affairs* 8:3 (2005)
- Eshleman, H. Frank. *Lancaster County Indians* (Self-published, 1909)
- Federal Writers' Project. *Pennsylvania: a Guide to the Keystone State* (University of Pennsylvania, 1940)
- Fiedel, Dorothy Burtz. *Haunted Lancaster County, Pennsylvania* (Science, 1994)
- Franklin, Benjamin. *The Conduct of the Paxton-Men, Impartially Represented: With Some Remarks On the Narrative* (Andrew Steuart, 1764).
- Gerhard, Ken and Nick Redfern. *Monsters of Texas* (CFZ, 2010)
- Godfrey, Linda S. *Hunting the American Werewolf* (Trails, 2006)

-----. *The Michigan Dogman* (Unexplained Research Publishing Company, 2010)
- Gordon, Stan. "Strange creature reported in Indiana County." *The Anomalies Zone* 3:1 (Fall/Winter 1996)

-----. *Really Mysterious Pennsylvania: UFOs, Bigfoot & Other Weird Encounters, Casebook One* (Stan Gordon, 2010)
- Griffin, James B., *et. al.* "A mammoth fraud in Science." *American Antiquity* 53:3 (July 1988)
- Grimm, Jakob and James Steven Stallybrass (translator). *Teutonic Mythology* (George Bell & Sons, 1883)
- Groff, Walter S. *The Groff Book, vol. 1* (Douglas W. Groff, 1985)
- Guinard, Reverend Joseph E. "Witiko among the Tete-de-Boule." *Primitive Man* 3:3-4 (July-October, 1930)
- Hagen, Tony J. "Fisherman hooks piranha in Delaware." *NJ.com* (posted: September 24, 2009): http://www.nj.com/mercer/index.ssf/2009/09/fisherman_hooks_piranha_in_del.html
- Hall, Mark A. *Natural Mysteries, Second Edition.* (Mark A. Hall Publications, 1991)
- Heinrichs, Allison M. "Alligator – free to a good home." *Pittsburgh Post-Gazette* (April 23, 2002)
- Heltzel, Bill. "State seizes exotic bearcat." *Pittsburgh Post-Gazette* (April 23, 2002)
- Hoffman, W.J. "Folklore of the Pennsylvania Germans, Part I." *Journal of American Folklore* 1:2 (1888)

-----. "Folklore of the Pennsylvania Germans, Part II." *Journal of American Folklore* 2:4 (1889)

-----. "Popular superstitions." *Appleton's Popular Science Monthly, Volume I*, William Jay Youmans, ed. (1896-1897)
- Ieraci, Ron. *Pennsylvania Haunts & History.* http://sites.google.com/site/hauntsandhistory/pennslvaniahaunts&history
- Johnston, Basil. *The Manitous* (Minnesota Historical Society Press, 2001)
- Johnston, George. *History of Cecil County, Maryland* (Self-published, 1881)

- Jordan, John W., ed. *A History of Delaware County, Pennsylvania and Its People* (Lewis Historical Publishing Company, 1914)
- Keel, John A. *The Mothman Prophecies* (Tor, 2002 ed.)
- Korson, George G. *Black Rock: Mining Folklore of the Pennsylvania Dutch* (Johns Hopkins University Press, 1960)
- Kriebel, David W. *Powwowing Among the Pennsylvania Dutch: A Traditional Medical Practice in the Modern World* (Pennsylvania State University Press, 2007)
- Lake, Matt, Mark Sceurman and Mark Moran. *Weird Pennsylvania* (Sterling Publishing Company, 2005)
- Lankford, George E. "Pleistocene animals in folk memory." *Journal of American Folklore* 93 (1980)
- Leighton, Alexander H. and Jane M. Hughes. "Cultures as a causative of mental disorder." *Milbank Memorial Fund Quarterly* 39:3 (1961)
- Linn, John B. *History of Centre and Clinton Counties, Pennsylvania* (L.H. Everts, 1883)
- Lyman, Robert R. *Forbidden Land: Strange Events in the Black Forest, Volume I* (Leader, 1971)
- -----. *Amazing Indeed: Strange Events in the Black Forest, Volume II* (Potter Enterprise, 1973)
- Maberry, Jonathan. "When sane people see weird things." *Jonathan Maberry's Big Scary Blog* (December 2, 2007). http://jonathanmaberry.blogspot.com/2007/12/when-sane-people-see-weird-things.html
- Maberry, Jonathan and David F. Kramer. *They Bite: Endless Cravings of Supernatural Predators* (Citadel, 2009)
- Masterson, Teresa. "Growling cougar watches man, dog near Del. townhouses." *NBC Philadelphia* (posted November 10, 2010): http://www.nbcphiladelphia.com/news/local-beat/Growling-Cougar-Watches-Man-Dog-Near-Del-Townhouses.html
- McCloy, James F. and Ray Miller, Jr. *The Jersey Devil* (Middle Atlantic Press, 1976)
- -----. *Phantom of the Pines: More Tales of the Jersey Devil* (Middle Atlantic Press, 1998)
- McGinnis, J. Ross. *Trials of Hex* (Davis/Trinity, 2000)
- Mercer, Henry C. *The Lenape Stone, or the Indian and the Mammoth* (G.P. Putnam's Sons, 1885)
- Merritt, Joseph F. *Guide to the Mammals of Pennsylvania* (University of Pittsburgh Press, 1987)
- Meurger, Michel. "The lindorms of Småland." *Arv: Nordic Yearbook of Folklore* 52 (1996)
- Mickle, Isaac. *Reminiscences of Old Gloucester: or Incidents in the History of the Counties of Gloucester, Atlantic and Camden* (Townsend Ward, 1845)
- Miller, Ernest C. "Place names in Warren County, Pennsylvania." http://www.warrenhistory.org/Places%20and%20Names%20by%20Ernie20Miller.html
- Miskelly, Colin M. "The identity of the hakawai." *Notornis* 34 (1987)
- Moore, John L. "The 'Piasa' as a representation of the 'underwater panther'." *The Cryptozoology Review* 3:1 (summer 1998).

- Moskowitz, Clara. "Ancient amphibian skull discovered at Pittsburgh airport." *FoxNews.com.* http://www.foxnews.com/scitech/2010/03/15/ancient-amphibian-skull-discovered-airport/
- Murray, Louise Welles. "Aboriginal sites in and near 'Teaoga', now Athens, Pennsylvania." *American Anthropologist* 23:2 (April-June 1921)
- Musinsky, Gerald. "Return of the thunderbird: avian mystery of the Black Forest." *Fate* (November 1995)
- Nesbitt, Mark. *Ghosts of Gettysburg* (Thomas, 1991)
- Noon, Mark A. *Yuengling: A History of America's Oldest Brewery* (McFarland & Company, 2005)
- Opsasnick, Mark. *The Maryland Bigfoot Digest* (Xlibris, 2004)
- Parker, L.A. "Don't be sure it's safe to go back in the water." *The Trentonian* (September 3, 2009)
- Pickel, Janet. "Alligator pulled out of Schuylkill River." http://www.pennlive.com/midstate/index.ssf/2008/10/alligator_pulled_out_of_schuyl.html.
- Polec, Don. "A glimpse of the 'Yardley Yeti'?" http://abclocal.go.com/wpvi/story?section=resources&id=4626922
- Prosek, James. *Eels: An Exploration, from New Zealand to the Sargasso, of the World's Most Mysterious Fish* (HarperCollins, 2010)
- Redfern, Nick. *Man-Monkey: In Search of the British Bigfoot* (CFZ, 2007)
- Reilly, P.J. "South American pacu hooked in Conestoga." *Lancaster Intelligencer Journal/New Era* (August 31, 2009)
- Rhoads, Samuel N. *The Mammals of Pennsylvania and New Jersey* (Philadelphia, 1903)
- Rose, Carol. *Giants, Monsters & Dragons* (W.W. Norton, 2000)
- Rosén, Sven. "The dragons of Sweden." *Fate* (April 1982)
- Schiavo, Christine. "Wallaby bounces back into captivity." *Philadelphia Inquirer* (July 7, 2004)
- Seibold, David J. and Charles J. Adams III. *Pocono Ghosts, Legends and Lore* (Exeter House, 1991)
- Shoemaker, Henry W. *Pennsylvania Mountain Stories* (Reading Times Publishing Company, 1909)

-----. *Black Forest Souvenirs* (Bright-Faust, 1914)
-----. *Extinct Pennsylvania Animals, Volume I* (Altoona Tribune Company, 1917)
-----. *Extinct Pennsylvania Animals, Volume II* (Altoona Tribune Company, 1919)
-----. *North Mountain Mementos* (Altoona Tribune Company, 1920)
-----. *South Mountain Sketches* (Altoona Tribune Company, 1920)
-----. *The Black Bear of Pennsylvania (Ursus americanus)* (Altoona Tribune Company, 1921)
-----. *Felis Catus in Pennsylvania?* (Altoona Tribune Company, 1922)

- Siebert, F.T. "Mammoth or stiff-legged bear." *American Anthropologist* 39 (1937)
- Sikes, Wirt. *British Goblins* (Sampson Low, Marston, Searle & Rivington, 1880)
- Skinner, Charles M. *Myths & Legends of Our Own Land* (2 vols.) (J.B. Lippincott & Company, 1896)

-----. *American Myths & Legends* (2 vols.) (J.B. Lippincott & Company, 1903)

- Smith, John. *Captain John Smith: Writings with Other Narratives of Roanoke, Jamestown, and the First English Settlement of America* (Library of America, 2007)
- Steadman, D.W. and N.G. Miller. "California condor associated with spruce-jack pine woodland in the late Pleistocene of New York." *Quaternary Research* 28 (1987)
- Strong, W.D. "North American traditions suggesting a knowledge of the mammoth." *American Anthropologist* 36 (1934)
- Summers, Montague. *The Werewolf in Lore and Legend* (Dover, 2004)
- Swauger, James L. "The Parker's Landing petroglyphs site, 26CL1." *Pennsylvania Archaeologist* 1-2 (1966).
- Swope, Robin. "Something in the blizzard." *Pittsburgh paranormal Examiner* (http://www.examiner.com/paranormal-in-pittsburgh/something-the-blizzard)
- -----. "The giant mound builders of Erie." *Pittsburgh Paranormal Examiner* (http://www.examiner.com/paranormal-in-pittsburgh/the-giant-mound-builders-of-erie)
- -----. "The Lake Erie 'storm hag,' demonic siren of the Great Lakes." *Pittsburgh Paranormal Examiner* (http://www.examiner.com/paranormal-in-pittsburgh/the-lake-erie-storm-hag-demonic-siren-of-the-great-lakes)
- -----. "The thing that moves at night." *Pittsburgh Paranormal Examiner* (http://www.examiner.com/paranormal-in-pittsburgh/the-thing-that-moves-at-night)
- "The outsider's view of the Mennonite Church." http://www.bfchistory.org/kulp2.htm
- Thwaites, Reuben Gold, ed. *Early Western Travels 1748-1846, Volume I* (Arthur H. Clark, 1904)
- Trevelyan, Marie. *Folk-Lore and Folk-Stories of Wales* (E.P. Publishing, 1973)
- Tryon, Henry H. *Fearsome Critters* (Idlewild, 1939)
- Waldron, David and Christopher Reeve. *Shock! The Black Dog of Bungay* (Hidden Publishing, 2010)
- Walls, Alyson. "Animal on front porch was a... well, no one seems to know." *Beaver County Times* (April 22, 2002)
- Watson, John F. *Historic Tales of Olden Times, Concerning the Early Settlement and Progress of Philadelphia and Pennsylvania* (E, Littell, 1833)
- Weslager, C.A. *The Delaware Indians: A History* (Rutgers University Press, 1972)
- -----. *New Sweden on the Delaware, 1638-1655* (Middle Atlantic Press, 1988)
- WGAL (Lancaster). "IMAGES: hairless coyote spotted." http://www.wgal.com/slideshow/news/24323576/detail.html
- Wheeler, Tim. "At loggerheads: rare sea turtle sighting in Bay." *B'more Green Blog* (posted August 14, 2009): http://weblogs.baltimoresun.com/features/green/2009/08 at_loggerheads_rare_sea_turtle.html
- White, Thomas. *Forgotten Tales of Pennsylvania* (History Press, 2009)
- Wilson, Patty A. *Totally Bizarre Pennsylvania* (Piney Creek Press, 2008)
- -----. *Monsters of Pennsylvania* (Stackpole Books, 2010)
- WPXI (Pittsburgh). "Alligator blamed for starting Lawrence County building fire." http://www.wpxi.com/news/18863199/detail.html
- Wysong, Thomas Turner. *The Rocks of Deer Creek, Harford County, Maryland* (A.J. Conlon, 1880)

## OTHER SOURCES

- "Common myna." *Birds of North America.* http://bna.birds.cornell.edu/bna/species/583/articles/introduction
- "Death-omens in Wales." *Frank Leslie's Popular Monthly, vol. 22* (Frank Leslie's Publishing House, 1886)
- "Delaware cougar confirmations." *Eastern Cougar Network* (posted November 26, 2002):
- http://www.easterncougarnet.org/delaware_cougar_confirmations.htm
- "Le loup garou." *Werewolves.* http://www.werewolves.com/le-loup-garou/
- "Michaux State Forest public use map." *Pennsylvania Department of Conservation and Natural Resources.* (http://www.dcnr.state.pa.us/ucmprd1/groups/public/documents/document D_000828.pdf)
- "Notable inmates – Morris Bolber." *Eastern State Penitentiary* (http://www.easternstate.org/learn/notable-inmates)
- "Pa. residents see wayward wallaby." http://cbs11tv.com/pets/Wallaby.Fleetwood.Pennsylvania.2.498968.html (January 23, 2007)

# STILL ON THE TRACK OF UNKNOWN ANIMALS

T he Centre for Fortean Zoology, or CFZ, is a non profit-making organisation founded in 1992 with the aim of being a clearing house for information, and coordinating research into mystery animals around the world.

We also study out of place animals, rare and aberrant animal behaviour, and Zooform Phenomena; little-understood "things" that appear to be animals, but which are in fact nothing of the sort, and not even alive (at least in the way we understand the term).

Not only are we the biggest organisation of our type in the world, but - or so we like to think - we are the best. We are certainly the only truly global cryptozoological research organisation, and we carry out our investigations using a strictly scientific set of guidelines. We are expanding all the time and looking to recruit new members to help us in our research into mysterious animals and strange creatures across the globe.

Why should you join us? Because, if you are genuinely interested in trying to solve the last great mysteries of Mother Nature, there is nobody better than us with whom to do it.

Members get a four-issue subscription to our journal *Animals & Men*. Each issue contains nearly 100 pages packed with news, articles, letters, research papers, field reports, and even a gossip column! The magazine is Royal Octavo in format with a full colour cover. You also have access to one of the world's largest collections of resource material dealing with cryptozoology and allied disciplines, and people from the CFZ membership regularly take part in fieldwork and expeditions around the world.

The CFZ is managed by a three-man board of trustees, with a non-profit making trust registered with HM Government Stamp Office. The board of trustees is supported by a Permanent Directorate of full and part-time staff, and advised by a Consultancy Board of specialists - many of whom are world-renowned experts in their particular field. We have regional representatives across the UK, the USA, and many other parts of the world, and are affiliated with other organisations whose aims and protocols mirror our own.

You'll find that the people at the CFZ are friendly and approachable. We have a thriving forum on the website which is the hub of an ever-growing electronic community. You will soon find your feet. Many members of the CFZ Permanent Directorate started off as ordinary members, and now work full-time chasing monsters around the world.

Write to us, e-mail us, or telephone us. The list of future projects on the website is not exhaustive. If you have a good idea for an investigation, please tell us. We may well be able to help.

We are always looking for volunteers to join us. If you see a project that interests you, do not hesitate to get in touch with us. Under certain circumstances we can help provide funding for your trip. If you look on the future projects section of the website, you can see some of the projects that we have pencilled in for the next few years.

In 2003 and 2004 we sent three-man expeditions to Sumatra looking for Orang-Pendek - a semi-legendary bipedal ape. The same three went to Mongolia in 2005. All three members started off merely subscribers to the CFZ magazine. Next time it could be you!

We have no magic sources of income. All our funds come from donations, membership fees, and sales of our publications and merchandise. We are always looking for corporate sponsorship, and other sources of revenue. If you have any ideas for fund-raising please let us know.

However, unlike other cryptozoological organisations in the past, we do not live in an intellectual ivory tower. We are not afraid to get our hands dirty, and furthermore we are not one of those organisations where the membership have to raise money so that a privileged few can go on expensive foreign trips. Our research teams, both in the UK and abroad, consist of a mixture of experienced and inexperienced personnel. We are truly a community, and work on the premise that the benefits of CFZ membership are open to all.

Reports of our investigations are published on our website as soon as they are available. Preliminary reports are posted within days of the project finishing.

Each year we publish a 200 page yearbook containing research papers and expedition reports too long to be printed in the journal. We freely circulate our information to anybody who asks for it.

We have a thriving YouTube channel, CFZtv, which has well over two hundred self-made documentaries, lecture appearances, and episodes of our monthly webTV show. We have a daily online magazine, which has over a million hits each year.

Each year since 2000 we have held our annual convention - the Weird Weekend. It is three days of lectures, workshops, and excursions. But most importantly it is a chance for members of the CFZ to meet each other, and to talk with the members of the permanent directorate in a relaxed and informal setting and preferably with a pint of beer in one hand. Since 2006 - the Weird Weekend has been bigger and better and held on the third weekend in August in the idyllic rural location of Woolsery in North Devon.

Since relocating to North Devon in 2005 we have become ever more closely involved with other community organisations, and we hope that this trend will continue. We have also worked closely with Police Forces across the UK as consultants for animal mutilation cases, and we intend to forge closer links with the coastguard and other community services. We want to work closely with those who regularly travel into the Bristol Channel, so that if the recent trend of exotic animal visitors to our coastal waters continues, we can be out there as soon as possible.

Apart from having been the only Fortean Zoological organisation in the world to have consistently published material on all aspects of the subject for over a decade, we have achieved the following concrete results:

• Disproved the myth relating to the headless so-called sea-serpent carcass of Durgan beach in Cornwall 1975
• Disproved the story of the 1988 puma skull of

Lustleigh Cleave

- Carried out the only in-depth research ever into the mythos of the Cornish Owlman.
- Made the first records of a tropical species of lamprey
- Made the first records of a luminous cave gnat larva in Thailand
- Discovered a possible new species of British mammal - the beech marten
- In 1994-6 carried out the first archival fortean zoological survey of Hong Kong
- In the year 2000, CFZ theories were confirmed when a new species of lizard was added to the British List
- Identified the monster of Martin Mere in Lancashire as a giant wels catfish
- Expanded the known range of Armitage's skink in the Gambia by 80%
- Obtained photographic evidence of the remains of Europe's largest known pike
- Carried out the first ever in-depth study of the ninki-nanka
- Carried out the first attempt to breed Puerto Rican cave snails in captivity
- Were the first European explorers to visit the `lost valley` in Sumatra
- Published the first ever evidence for a new tribe of pygmies in Guyana
- Published the first evidence for a new species of caiman in Guyana
- Filmed unknown creatures

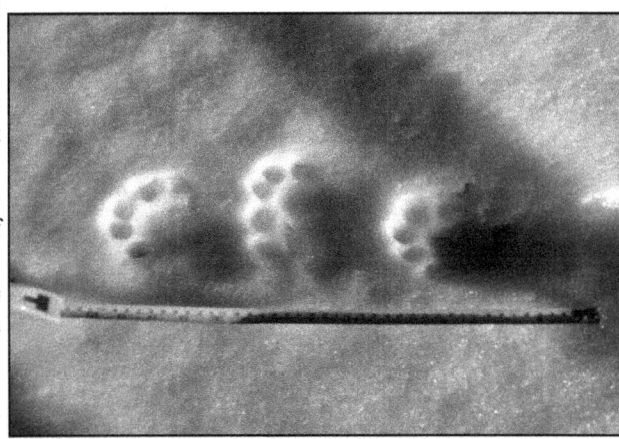

on a monster-haunted lake in Ireland for the first time
- Had a sighting of orang pendek in Sumatra in 2009
- Found leopard hair, subsequently identified by DNA analysis, from rural North Devon in 2010
- Brought back hairs which appear to be from an unknown primate in Sumatra
- Published some of the best evidence ever for the almasty in southern Russia

CFZ Expeditions and Investigations include:

- 1998 Puerto Rico, Florida, Mexico (Chupacabras)
- 1999 Nevada (Bigfoot)
- 2000 Thailand (Naga)
- 2002 Martin Mere (Giant catfish)
- 2002 Cleveland (Wallaby mutilation)

- 2003 Bolam Lake (BHM Reports)
- 2003 Sumatra (Orang Pendek)
- 2003 Texas (Bigfoot; giant snapping turtles)
- 2004 Sumatra (Orang Pendek; cigau, a sabre-toothed cat)
- 2004 Illinois (Black panthers; cicada swarm)
- 2004 Texas (Mystery blue dog)
- Loch Morar (Monster)
- 2004 Puerto Rico (Chupacabras; carnivorous cave snails)
- 2005 Belize (Affiliate expedition for hairy dwarfs)
- 2005 Loch Ness (Monster)
- 2005 Mongolia (Allghoi Khorkhoi aka Mongolian death worm)

- 2006 Gambia (Gambo - Gambian sea monster, Ninki Nanka and Armitage's skink
- 2006 Llangorse Lake (Giant pike, giant eels)
- 2006 Windermere (Giant eels)
- 2007 Coniston Water (Giant eels)
- 2007 Guyana (Giant anaconda, didi, water tiger)
- 2008 Russia (Almasty)
- 2009 Sumatra (Orang pendek)
- 2009 Republic of Ireland (Lake Monster)
- 2010 Texas (Blue Dogs)
- 2010 India (Mande Burung)
- 2011 Sumatra (Orang-pendek)

For details of current membership fees, current expeditions and investigations, and voluntary posts within the CFZ that need your help, please do not hesitate to contact us.

The Centre for Fortean Zoology,
Myrtle Cottage,
Woolfardisworthy,
Bideford, North Devon
EX39 5QR

Telephone 01237 431413
Fax+44 (0)7006-074-925
**eMail** info@cfz.org.uk

**Websites:**

www.cfz.org.uk
www.weirdweekend.org

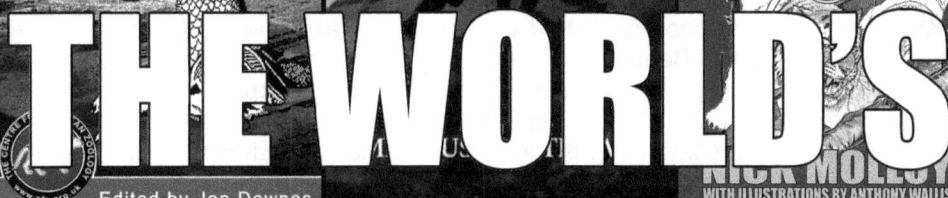

# THE WORLD'S WEIRDEST PUBLISHING COMPANY

# HOW TO START A PUBLISHING EMPIRE

Unlike most mainstream publishers, we have a non-commercial remit, and our mission statement claims that "we publish books because they deserve to be published, not because we think that we can make money out of them". Our motto is the Latin Tag *Pro bona causa facimus* (we do it for good reason), a slogan taken from a children's book *The Case of the Silver Egg* by the late Desmond Skirrow.

WIKIPEDIA: "The first book published was in 1988. *Take this Brother may it Serve you Well* was a guide to Beatles bootlegs by Jonathan Downes. It sold quite well, but was hampered by very poor production values, being photocopied, and held together by a plastic clip binder. In 1988 A5 clip binders were hard to get hold of, so the publishers took A4 binders and cut them in half with a hacksaw. It now reaches surprisingly high prices second hand.

The production quality improved slightly over the years, and after 1999 all the books produced were ringbound with laminated colour covers. In 2004, however, they signed an agreement with Lightning Source, and all books are now produced perfect bound, with full colour covers."

Until 2010 all our books, the majority of which are/were on the subject of mystery animals and allied disciplines, were published by `CFZ Press`, the publishing arm of the Centre for Fortean Zoology (CFZ), and we urged our readers and followers to draw a discreet veil over the books that we published that were completely off topic to the CFZ.

However, in 2010 we decided that enough was enough and launched a second imprint, `Fortean Words` which aims to cover a wide range of non animal-related esoteric subjects. Other imprints will be launched as and when we feel like it, however the basic ethos of the company remains the same: Our job is to publish books and magazines that we feel are worth publishing, whether or not they are going to sell. Money is, after all - as my dear old Mama once told me - a rather vulgar subject, and she would be rolling in her grave if she thought that her eldest son was somehow in `trade`.

Luckily, so far our tastes have turned out not to be that rarified after all, and we have sold far more books than anyone ever thought that we would, so there is a moral in there somewhere…

Jon Downes,
Woolsery, North Devon
July 2010

# CFZ PRESS

## Other Books in Print

*ORANG PENDEK: Sumatra's Forgotten Ape* by Richard Freeman
*THE MYSTERY ANIMALS OF THE BRITISH ISLES: London* by Neil Arnold
*CFZ EXPEDITION REPORT: India 2010* by Richard Freeman *et al*
*The Cryptid Creatures of Florida* by Scott Marlow
*Dead of Night* by Lee Walker
*The Mystery Animals of the British Isles: The Northern Isles* by Glen Vaudrey
*THE MYSTERY ANIMALS OF THE BRTISH ISLES: Gloucestershire and Worcestershire* by Paul Williams
*When Bigfoot Attacks* by Michael Newton
*Weird Waters – The Mystery Animals of Scandinavia: Lake and Sea Monsters* by Lars Thomas
*The Inhumanoids* by Barton Nunnelly
*Monstrum! A Wizard's Tale* by Tony "Doc" Shiels
*CFZ Yearbook 2011* edited by Jonathan Downes
*Karl Shuker's Alien Zoo* by Shuker, Dr Karl P.N
*Tetrapod Zoology Book One* by Naish, Dr Darren
*The Mystery Animals of Ireland* by Gary Cunningham and Ronan Coghlan
*Monsters of Texas* by Gerhard, Ken
*The Great Yokai Encyclopaedia* by Freeman, Richard
*NEW HORIZONS:* Animals & Men *issues 16-20 Collected Editions Vol. 4* by Downes, Jonathan
*A Daintree Diary -*
*Tales from Travels to the Daintree Rainforest in tropical north Queensland, Australia* by Portman, Carl
*Strangely Strange but Oddly Normal* by Roberts, Andy
*Centre for Fortean Zoology Yearbook 2010* by Downes, Jonathan
*Predator Deathmatch* by Molloy, Nick
*Star Steeds and other Dreams* by Shuker, Karl
*CHINA: A Yellow Peril?* by Muirhead, Richard
*Mystery Animals of the British Isles: The Western Isles* by Vaudrey, Glen
*Giant Snakes - Unravelling the coils of mystery* by Newton, Michael
*Mystery Animals of the British Isles: Kent* by Arnold, Neil

*Centre for Fortean Zoology Yearbook 2009* by Downes, Jonathan
*CFZ EXPEDITION REPORT: Russia 2008* by Richard Freeman *et al*, Shuker, Karl (fwd)
*Dinosaurs and other Prehistoric Animals on Stamps - A Worldwide catalogue* by Shuker, Karl P. N
*Dr Shuker's Casebook* by Shuker, Karl P.N
*The Island of Paradise - chupacabra UFO crash retrievals, and accelerated evolution on the island of Puerto Rico* by Downes, Jonathan
*The Mystery Animals of the British Isles: Northumberland and Tyneside* by Hallowell, Michael J
*Centre for Fortean Zoology Yearbook 1997* by Downes, Jonathan (Ed)
*Centre for Fortean Zoology Yearbook 2002* by Downes, Jonathan (Ed)
*Centre for Fortean Zoology Yearbook 2000/1* by Downes, Jonathan (Ed)
*Centre for Fortean Zoology Yearbook 1998* by Downes, Jonathan (Ed)
*Centre for Fortean Zoology Yearbook 2003* by Downes, Jonathan (Ed)
*In the wake of Bernard Heuvelmans* by Woodley, Michael A
*CFZ EXPEDITION REPORT: Guyana 2007* by Richard Freeman *et al*, Shuker, Karl (fwd)
*Centre for Fortean Zoology Yearbook 1999* by Downes, Jonathan (Ed)
*Big Cats in Britain Yearbook 2008* by Fraser, Mark (Ed)
*Centre for Fortean Zoology Yearbook 1996* by Downes, Jonathan (Ed)
*THE CALL OF THE WILD - Animals & Men issues 11-15 Collected Editions Vol. 3* by Downes, Jonathan (ed)
*Ethna's Journal* by Downes, C N
*Centre for Fortean Zoology Yearbook 2008* by Downes, J (Ed)
*DARK DORSET -Calendar Custome* by Newland, Robert J
*Extraordinary Animals Revisited* by Shuker, Karl
*MAN-MONKEY - In Search of the British Bigfoot* by Redfern, Nick
*Dark Dorset Tales of Mystery, Wonder and Terror* by Newland, Robert J and Mark North
*Big Cats Loose in Britain* by Matthews, Marcus
*MONSTER! - The A-Z of Zooform Phenomena* by Arnold, Neil
*The Centre for Fortean Zoology 2004 Yearbook* by Downes, Jonathan (Ed)
*The Centre for Fortean Zoology 2007 Yearbook* by Downes, Jonathan (Ed)
*CAT FLAPS! Northern Mystery Cats* by Roberts, Andy
*Big Cats in Britain Yearbook 2007* by Fraser, Mark (Ed)
*BIG BIRD! - Modern sightings of Flying Monsters* by Gerhard, Ken
*THE NUMBER OF THE BEAST - Animals & Men issues 6-10 Collected Editions Vol. 1* by Downes, Jonathan (Ed)
*IN THE BEGINNING - Animals & Men issues 1-5 Collected Editions Vol. 1* by Downes, Jonathan
*STRENGTH THROUGH KOI - They saved Hitler's Koi and other stories* by Downes, Jonathan
*The Smaller Mystery Carnivores of the Westcountry* by Downes, Jonathan
*CFZ EXPEDITION REPORT: Gambia 2006* by Richard Freeman *et al*, Shuker, Karl (fwd)
*The Owlman and Others* by Jonathan Downes
*The Blackdown Mystery* by Downes, Jonathan
*Big Cats in Britain Yearbook 2006* by Fraser, Mark (Ed)
*Fragrant Harbours - Distant Rivers* by Downes, John T

*Only Fools and Goatsuckers* by Downes, Jonathan
*Monster of the Mere* by Jonathan Downes
*Dragons:More than a Myth* by Freeman, Richard Alan
*Granfer's Bible Stories* by Downes, John Tweddell
*Monster Hunter* by Downes, Jonathan

# Fortean Words

The Centre for Fortean Zoology has for several years led the field in Fortean publishing. CFZ Press is the only publishing company specialising in books on monsters and mystery animals. CFZ Press has published more books on this subject than any other company in history and has attracted such well known authors as Andy Roberts, Nick Redfern, Michael Newton, Dr Karl Shuker, Neil Arnold, Dr Darren Naish, Jon Downes, Ken Gerhard and Richard Freeman.

Now CFZ Press are launching a new imprint. Fortean Words is a new line of books dealing with Fortean subjects other than cryptozoology, which is - after all - the subject the CFZ are best known for. Fortean Words is being launched with a spectacular multi-volume series called *Haunted Skies* which covers British UFO sightings between 1940 and 2010. Former policeman John Hanson and his long-suffering partner Dawn Holloway have compiled a peerless library of sighting reports, many that have not been made public before.

Other books include a look at the Berwyn Mountains UFO case by renowned Fortean Andy Roberts and a series of forthcoming books by transatlantic researcher Nick Redfern. CFZ Press are dedicated to maintaining the fine quality of their works with Fortean Words. New authors tackling new subjects will always be encouraged, and we hope that our books will continue to be as ground-breaking and popular as ever.

*Haunted Skies Volume One 1940-1959* by John Hanson and Dawn Holloway
*Haunted Skies Volume Two 1960-1965* by John Hanson and Dawn Holloway
*Haunted Skies Volume Three 1965-1967* by John Hanson and Dawn Holloway
*Haunted Skies Volume Four 1968-1971* by John Hanson and Dawn Holloway
*Grave Concerns* by Kai Roberts

*Police and the Paranormal* by Andy Owens
*Dead of Night* by Lee Walker
*Space Girl Dead on Spaghetti Junction* - an anthology by Nick Redfern
*I Fort the Lore* - an anthology by Paul Screeton
*UFO Down - the Berwyn Mountains UFO Crash* by Andy Roberts

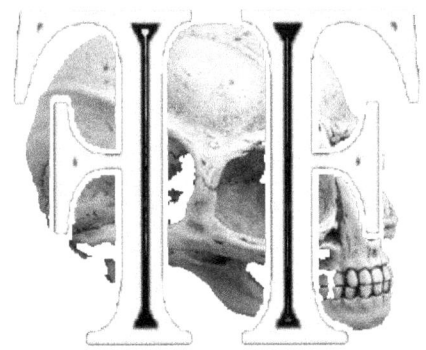

# Fortean Fiction

Just before Christmas 2011, we launched our third imprint, this time dedicated to - let's see if you guessed it from the title - fictional books with a Fortean or cryptozoological theme. We have published a few fictional books in the past, but now think that because of our rising reputation as publishers of quality Forteana, that a dedicated fiction imprint was the order of the day.

We launched with four titles:

*Green Unpleasant Land* by Richard Freeman
*Left Behind* by Harriet Wadham
*Dark Ness* by Tabitca Cope
*Snap!* By Steven Bredice

www.ingramcontent.com/pod-product-compliance
Lightning Source LLC
Chambersburg PA
CBHW070656100426
42735CB00039B/2165

i